PHYSICS

50 ESSENTIAL IDEAS

PHYSICS

50 ESSENTIAL IDEAS

ROBERT SNEDDEN

This edition published in 2023 by Sirius Publishing, a division of
Arcturus Publishing Limited,
26/27 Bickels Yard, 151–153 Bermondsey Street,
London SE1 3HA

ISBN: 978-1-3988-3025-7
AD010545UK

Printed in China

Contents

Introduction

'All science is either physics or stamp collecting.' So Ernest Rutherford, physicist and discoverer of the atomic nucleus is alleged to have remarked.

Through a combination of observation and experiment, science seeks to find the underlying rules that underpin reality. Physics is the science of matter and energy and the ways in which they interact to produce the universe. It is in many ways the science that lies at the heart of everything.

MATHS AND MEASUREMENT

Physics is largely concerned with things that can be measured and quantified. The most reliable way to discern meaningful patterns in the data gathered by experiment and observation is through mathematical analysis. Mathematics was key to opening up the physics of the universe. Johannes Kepler combined observational data with mathematics to show in 1609 that the planets moved around the sun in ellipses rather than circles as had been believed. Around 1670, Isaac Newton and Gottfried Leibniz independently developed a new theory of mathematics, calculus, that allowed physicists to precisely model and analyse dynamic systems. Data champions belief every time.

A MECHANICAL UNIVERSE

The universe for the physicists of the 18th and 19th centuries was a mechanistic one. They began to believe that it might be possible to quantify and understand everything. Laws were formulated to explain how things moved, how gases behaved, how heat was transferred from place to place. If only we could know the positions and velocities of all the objects in the universe and the forces acting on them, we would know where they would be at all future times. It didn't

Physics seeks to explain the laws that govern the universe.

really matter that this was practically impossible; theoretically at least it could be contemplated.

WORLD TURNED UPSIDE DOWN

The 20th century brought a profound challenge to these beliefs. Albert Einstein's relativity theories saw the speed of light dictate the very nature of time and space and gravity become a distortion in the new concept of spacetime. Quantum mechanics confronted what was meant by 'real' as certainty gave way to probability and potential in a strange new world where light could be simultaneously both a wave and a particle, and where the properties of an object had no real meaning until you measured them.

THE CHALLENGE OF PHYSICS

One of the greatest challenges thereafter for physicists was finding a way to reconcile these twin revolutions. Uniting Einstein's large-scale spacetime universe with the deep strangeness of the subatomic quantum realm in an overarching 'theory of everything' is a goal that continues to the present day but has so far proved elusive.

Physics doesn't stand still. It continues to make new discoveries – charting events in the first fractions of a second after the Big Bang that brought the universe into existence; contemplating what might go on inside a black hole; investigating the very structure of the atom and recreating the energy that powers the stars.

This book takes a look at just some of the stepping stones physics has used on the way to piecing together what Stephen Hawking called 'the grand design of the universe'. The journey is far from over.

1

The fact of the matter

What is everything made of?

The things that make up ourselves and our universe are many and varied. Some things we can touch and hold; other things seem insubstantial. Some things are obvious; some are undetectable without sophisticated instruments. All of these things are formed from matter.

Matter is anything in nature that has shape, form and mass. Pondering what was the most important knowledge we possess about the universe, physicist Richard Feynman decided that the answer was: 'All things are made of atoms – little particles that move around in perpetual motion, attracting each other when they are a little distance apart, but repelling upon being squeezed into one another.'

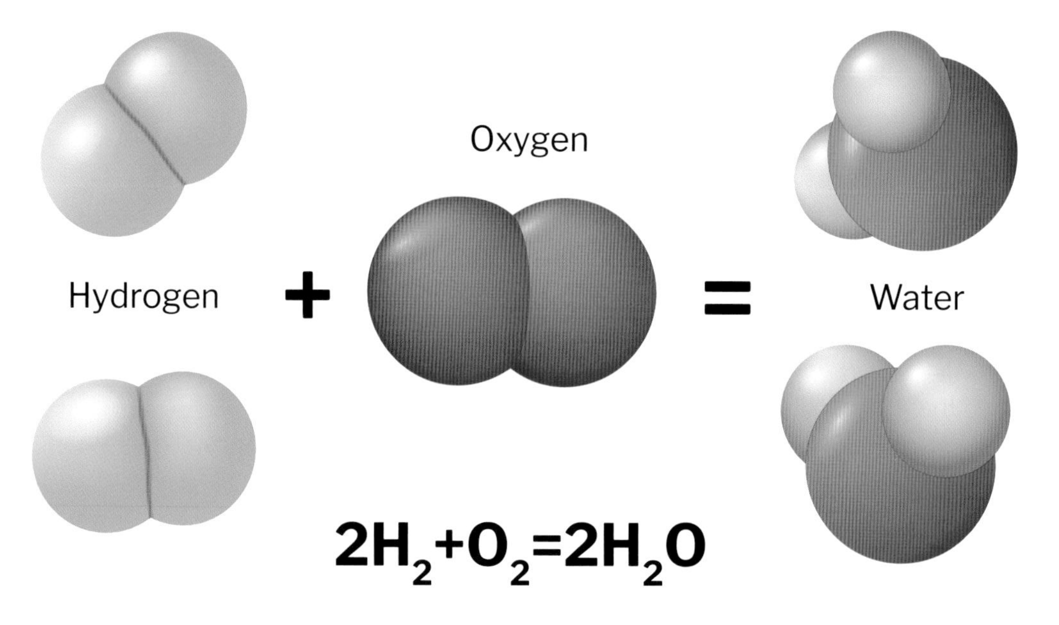

According to the atomic theory, each element is composed of a great number of atoms with unique characteristics. The atoms of different elements can combine to form compounds.

John Dalton.

ATOMIC THEORY

The idea of matter being formed from invisibly small, indivisible particles can be traced back to the 5th century BCE and Greek philosopher Democritus. He named these indivisible particles atomos, suggesting that they were eternal and indestructible, and that each material consisted of its own unique form of atomos. This remarkable insight into the nature of the physical universe was largely ignored for two millennia.

The atomic theory of matter was revived by physicist John Dalton at the beginning of the 19th century. Initial resistance soon gave way to an acceptance of the idea that all matter is composed of atoms of different weights that combine in simple ratios to form compounds. Nineteenth-century atomic theory also proposed that atoms were indestructible and that the atoms forming a particular element were all identical. Although we now know that atoms can indeed be split into smaller particles, for everyday practical purposes the atomic theory holds true.

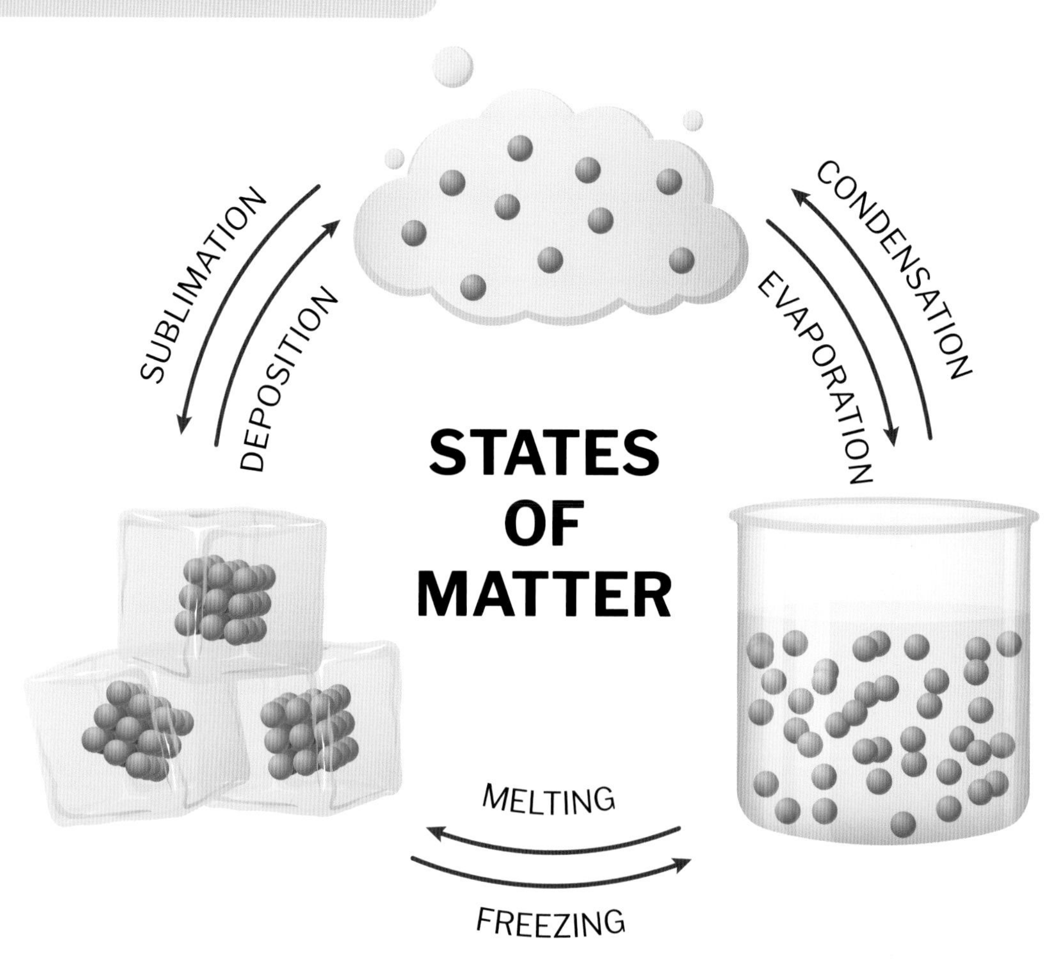

Matter can exist in three different states – solid, liquid and gas – and can transition from one to the other as a result of physical forces such as changes in temperature or pressure. The temperature and pressure at which gas, liquid and solid coexist in thermodynamic equilibrium is known as the triple point.

STATES OF MATTER – CHANGES OF STATE

All matter is composed of billions upon billions of tiny particles continually interacting with each other. The arrangement of these particles gives matter its form and properties. Solids, liquids and gases are three states of matter, each with its own characteristics. Changing from one state to another is a physical process – no chemical reactions are involved. When a substance changes from one state to another the particles it is made up of take up a new physical arrangement, but their number stays the same. The mass of the substance is conserved.

In a solid, the particles are held together in the firm grip of powerful forces. The particles in a liquid are bound by weaker forces and, although still bound together, they can flow, taking the shape of whatever container they are poured into. The forces holding a gas are weaker still. The particles in a gas can spread far apart from each, filling whatever container they are held in.

PARTICLES IN MOTION

The particles in a fluid (a liquid or gas) are in constant motion. They spread from areas of high concentration to areas of low concentration – a process called diffusion. This brings about the gradual mixing of different liquids and gases. Think about how a drop of ink will spread through a glass of water, or how the scent of a perfume can be detected at a distance from the person wearing it.

The way in which the particles in a substance move is dependent on their temperature. Heating an object causes the particles in it to gain kinetic energy, causing them to move faster. The particles in a solid are fixed in place but they are not motionless. They can be thought of as continually vibrating back and forth. As the solid is heated these vibrations become stronger, causing the particles to move a little further apart and the solid to expand – a phenomenon known as thermal expansion. Engineers must take thermal expansion into account, installing expansion joints in structures such as bridges that are subject to changes in temperature.

LATENT HEAT

Changing a substance from a solid to a liquid, or a liquid to a gas, requires energy. This energy is called latent heat. The energy needed to change a solid to a liquid is called the latent heat of fusion, that to change a liquid to a gas is the latent heat of vaporization. The latent heat of vaporization is much higher than the latent heat of fusion. When a substance melts or boils energy is being used to overcome the forces holding the particles together, so the temperature remains the same. The reverse is also true. When a liquid freezes, or a gas condenses, as bonds form energy is released, so again the temperature remains the same.

Internal energy and temperature

Temperature is a measure of the average kinetic energy of the particles in a substance – the faster they move, the higher the temperature. The internal energy of a substance is a measure of the total kinetic and chemical potential energy of the particles that make it up. A large, cold object can have more internal energy than a small, hot one because it has a greater number of particles.

2

Persistence of motion

Motion and inertia

For centuries the generally accepted ideas about motion were those of the ancient Greek philosopher Aristotle. He believed that an object will only move if pushed and once the pushing stops the object stops too. This didn't explain why a discus didn't immediately fall to the ground after it left the hand, or why an arrow continued to fly after it left the bow.

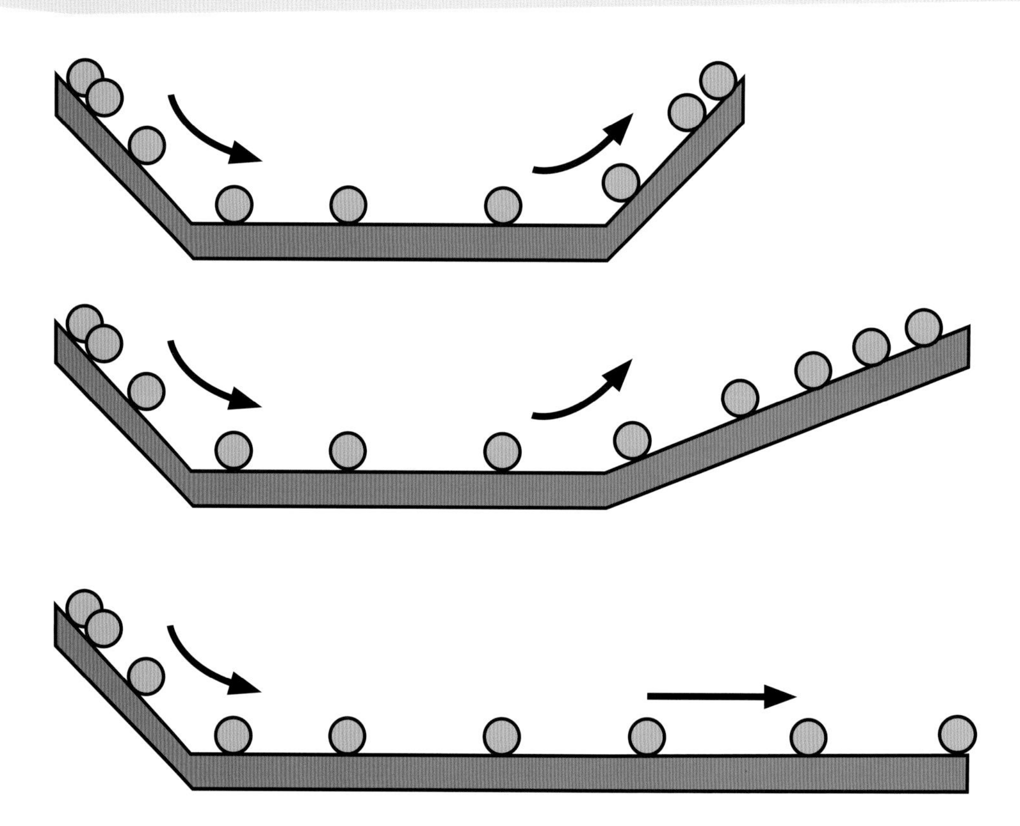

Galileo discovered that a ball rolled down one plane and up another will never reach a height greater than its initial height, whatever the slope of the planes. If the second plane is horizontal, the ball will roll forever unless stopped by an external force.

In the Middle Ages, scholars such as the Persian philosopher Avicenna (980–1037) asserted that when a body is set in motion by a force, it keeps moving due to what was termed its 'impetus'. The impetus imparted to the object by the initial force keeps the object in motion until another opposing force counteracts the impetus. Avicenna thought of this impetus as an internal force, actively pushing the object forward.

Much later, in the 1630s, Italian mathematician and scientist Galileo Galilei was studying the motion of balls rolling on inclined planes. He observed that a ball rolled down one plane and up another would reach the same height on the second plane as its initial height on the first, no matter how steeply inclined his planes were. He reasoned from this that if the second plane was horizontal the ball would keep rolling forever unless something stopped it. Galileo reasoned that, provided no external forces were acting, there was no real difference between an object moving at a steady speed and direction and an object that wasn't moving at all. This was the basis of the principle of inertia.

THE FIRST LAW OF MOTION

Isaac Newton's laws of motion, published in 1687, set out the relationships between the forces acting on an object and its motion. The first of Newton's laws of motion is sometimes referred to as the

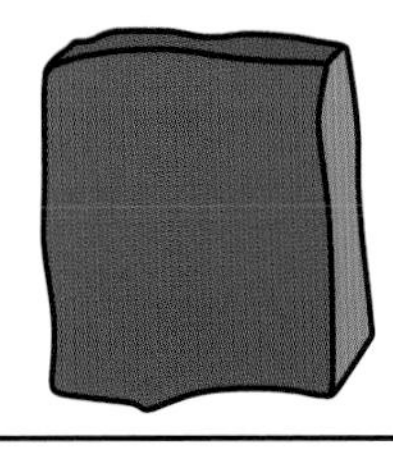

If an object is acted on by balanced forces, forces of equal magnitude acting in opposite directions, they cancel each other out and the object stays at rest. The object will only move when acted upon by unbalanced forces.

law of inertia. It states:

> *An object will remain at rest or continue to move in the same direction and at the same speed unless acted on by an unbalanced force.*

Every object has a property called inertia. This is the resistance it has to a change in its direction of motion or position. The inertia of an object is directly proportional to its mass: the more massive it is, the greater the force that must be exerted to get it moving or to change its direction if it is already moving. We know this from experience – it is easier to push a balloon than a boulder. All objects, whether moving or stationary, have inertia. If you are in a vehicle that brakes suddenly you will find yourself being jerked forward. This is because the force that acted to slow the vehicle did not act on you. Inertia means that you continue moving in the direction of travel until, hopefully, your motion is arrested by a seatbelt.

What is a force?

A force is defined as a push or pull exerted by one object on another that changes the speed, direction of movement, or the shape of the object. A force can require physical contact, such as the force of a hammer striking a nail, or be non-contact, such as the force of gravity holding a satellite in orbit.

When several forces act on an object at the same time their combined effects can be summed as a single force, called the resultant force. Balanced forces cancel each other out and do not change the object's motion. Unbalanced forces cause a change in motion.

The unit of force is the newton, named after Isaac Newton. It is defined as the force needed to accelerate a 1 kg (2.2 lb) object by 1 m/s (2.2 mph).

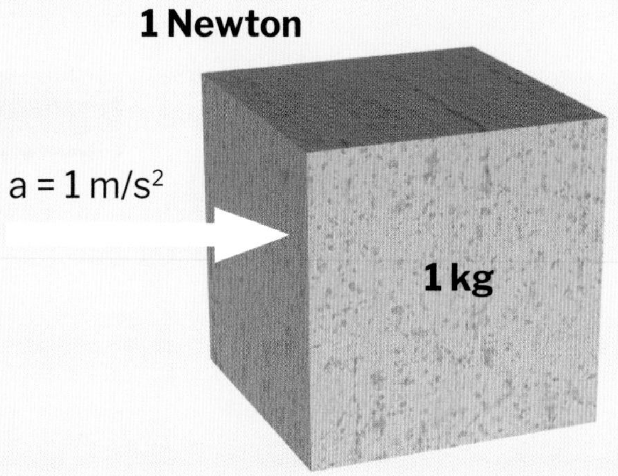

3

It's just rocket science

Action and reaction

One of Isaac Newton's insights into motion was that forces always act in pairs. Rocket science isn't complicated. In order to get off the ground you simply have to push hard enough in the opposite direction.

Newton's second law of motion is one of the most important of the physical laws. It describes how much an object accelerates when a net force is applied to it. It can be expressed simply as:

$F = ma$
(force equals mass multiplied by acceleration)

The larger the force that is applied, the greater the acceleration will be, and the greater the mass, the smaller the acceleration. Another way of looking at it is to say that the rate of change of the object's momentum, which as we have seen is a product of its mass and velocity, is proportional to the magnitude of the force applied.

Newton's third law of motion states that:

When two objects interact, the force acting on one object as the result

A colossal 3.4 million kilograms of thrust, equivalent to 160 million horsepower, lifts a fully fuelled 2.8 million-kilogram Saturn V rocket off the launch pad.

of this interaction will be equal and opposite to the force from the interaction acting on the other object

More succinctly:

For every action there is an equal and opposite reaction.

The conclusion from this is that forces act in pairs. If one object exerts a force on another, an equal and opposite force is exerted by the second object on the first. Newton's third law explains how a rocket engine generates thrust. The combustion of fuel in the engine produces hot exhaust gas that flows out through the nozzle of the engine. This produces an equal and opposite thrusting force on the rocket that accelerates it in accordance with the second law.

If the force increases so does the acceleration. As a rocket burns up its fuel its mass decreases and so, as the rate of change of momentum is proportional to the mass, then, so long as the thrust remains the same, the rate of acceleration will increase.

MOMENTUM

When an object in motion collides with another object the effect that one has on the other depends on a quantity known as momentum. Momentum is a product of the object's mass multiplied by its velocity. The greater the mass of the object and the faster it is moving the greater its momentum. The idea of momentum as product of mass and speed was introduced by 17th-century French philosopher René Descartes. Dutch

Mass calculations

Rearranging the second law equation gives us another important characteristic of an object, its inertial mass.

$m = F/a$

The inertial mass of an object is a measure of how difficult it is to get it moving or to change its velocity. If a known force is applied to an unknown mass and the resulting acceleration measured, the value for the mass can be accurately determined.

The gravitational mass of an object is determined by comparing the force of gravity of the unknown mass to the force of a known mass. Typically, this would be done using some sort of balance. Interestingly, there is no difference between the inertial mass and the gravitational mass. Albert Einstein used this fact as the jumping off point for his theory of general relativity.

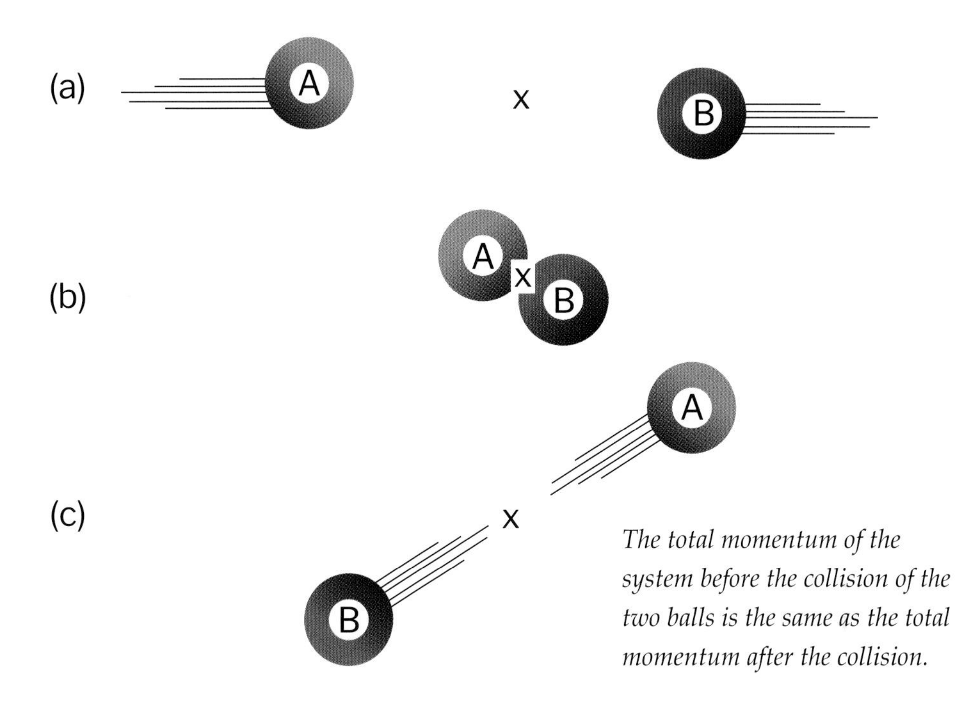

The total momentum of the system before the collision of the two balls is the same as the total momentum after the collision.

physicist Christiaan Huygens refined Descartes' thinking, suggesting that it was necessary to consider not only the object's speed, but also its direction, in other words, its velocity.

Momentum is traditionally denoted by the letter P, so the definition of momentum can be written as P = mv (where the object has mass m and is moving at velocity v). P and v are both vector quantities, that is they have magnitude and direction. One of the great laws of physics is that momentum is conserved. In a closed system, that is one that is not affected by any external forces:

total momentum before a collision = total momentum after the collision

The conservation of momentum explains why a firearm recoils when it is discharged. The projectile gains forward momentum and the firearm gains equivalent backward momentum. Before firing, the total momentum of the firearm / projectile system is zero. The total momentum of the projectile and firearm after the discharge is also zero as each is moving in an opposite direction.

If an object moving in one direction is considered to have positive momentum, then an object moving in the opposite direction would have negative momentum. A system consisting of two objects of equal mass moving together from opposite directions at the same speed has a total momentum of zero as the momentum of one cancels out the momentum of the other. When they collide, they come to a halt. The total momentum before the collision is the same as the total momentum afterwards – zero.

In terms of Newton's third law, suppose two objects, m1 and m2, interact, and that the forces acting on the objects as a result of the interaction are F1 and F2, respectively. According to the third law, F2 is opposite to F1, in other words equal to −F1, which means that F1 + F2 = 0 and that there is no resultant force acting on the system as a whole. Therefore, the total momentum of the system m1v1 + m2v2 will remain constant.

You may wonder why, if equal and opposite forces cancel out, it is ever possible to open a door that is pushing back at you just as hard as you are pushing it. The answer is that, in accordance with Newton's second law dealing with acceleration, only the forces acting on the door, or any other object, are considered, and not the force the door itself is exerting. Provided sufficient force is applied to get the door moving, but not enough for you to fall backwards, all will be well.

Newton derived his laws from three fundamental quantities that underpin all of science – time, mass and distance. Knowing the time an object takes to travel a set distance enables you to calculate its velocity (speed and direction). Mass tells you how much matter that object contains and therefore what amount of force you'll need to move it. Multiply the mass by the velocity and you'll get the object's momentum, which will tell you how hard it's going to be to get it moving, or to stop it.

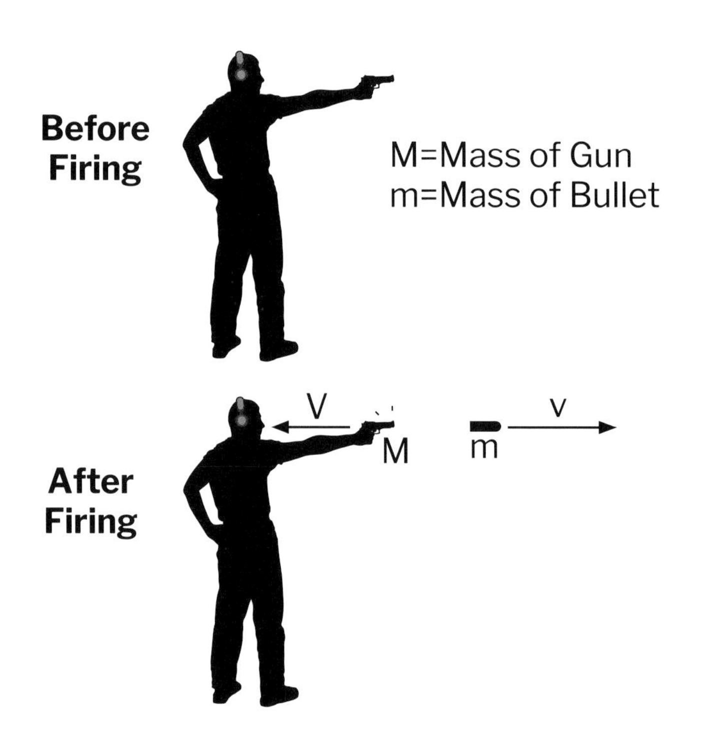

Before firing the gun, the net momentum is zero. Conservation of momentum means that the total momentum must be zero after firing as well. The momentum of the recoiling gun, MV, must be the same as the momentum of the bullet, mv. If a gun with a mass of 1.5 kg (3.3 lb) fires a bullet of mass 20 g (0.7 oz) at a velocity of 500 m/s, the recoil velocity of the gun will be 6.66 m/s.

4

Free falling

The law of gravity

The 4th-century BCE ideas of Aristotle about why objects fell to earth dominated thinking for centuries. He believed that heavy objects fell because they were returning to their 'natural place' and smoke rose because it belonged in the air. It wasn't until the 17th century that Galileo and Newton seriously challenged these notions.

Just as he had with his ideas about motion, Galileo also laid the groundwork for Newton's understanding of gravity. For centuries it had been assumed that a heavier object would fall faster than a lighter one. Galileo sought to prove or disprove this assumption by experiment. Rolling a ball down an inclined plane allowed him to slow the rate of acceleration sufficiently to be able to time it. Timing the rate of descent of differently weighted balls and considering the slowing effect of friction, Galileo concluded that freely falling objects accelerated uniformly at a rate that was independent of their mass. That a feather might fall more slowly than a rock was simply a consequence of the greater effect of air resistance on the former.

Galileo's discoveries set the groundwork for what is now referred to as classical mechanics, embracing the movement of falling bodies and projectiles. He laid a path for Newton to follow on the way to his laws of motion and the theory of gravity.

THE APPLE AND THE MOON

In popular legend, Newton's musings on gravity were inspired by his observation of a falling apple. True or not, we can imagine that his thinking went along these lines. There must, in accordance with the second law of motion, be a force that acts to accelerate the apple, since its velocity changes from zero attached to the tree to its final velocity when it hits the ground. Let's call this force 'gravity'. Now suppose that the apple tree is twice as tall. Will the apple still fall? Yes, it will, which suggests that the force of gravity extends at least to the top of the highest tree. Does it go further? If an apple falls to the earth, why doesn't the moon? It took real genius to conclude that the moon is falling (see overleaf).

Newton knew that any explanation he came up with to explain the motion of

The gravitational force acting between the earth and the moon is calculated by multiplying the masses (m1×m2), dividing by the square of the distance, r, between them and multiplying by the gravitational constant, G.

$$F = G\,\frac{m_1 \times m_2}{r^2}$$

F: the force between the masses
G: gravitational constant
r: distance
m: mass

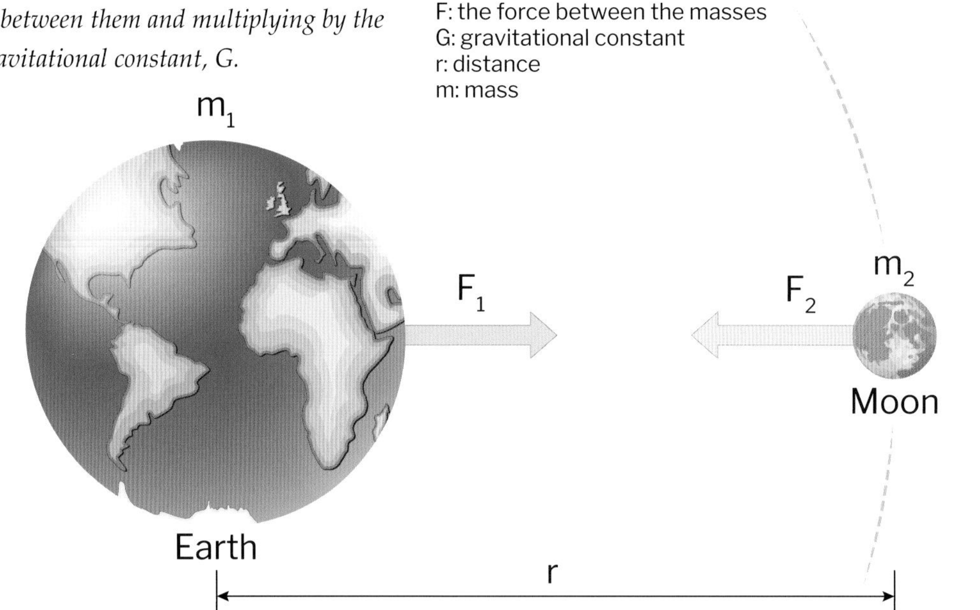

both the apple and the moon would have to encompass both Galileo's findings and those of the astronomer Johannes Kepler who had been painstakingly charting the paths of the planets around the sun. Newton determined that between any two objects there is always a gravitational force that attracts them to each other. The strength of the force depends on the masses of the objects and on the distance between them.

Gravity obeys an inverse square law, which means that the magnitude of the force decreases by the square of the distance. Therefore, if you double the distance between two objects the force that draws them together reduces to just a quarter of what it was. At five times the distance the force is reduced to a 25th. The law of universal gravitation, therefore, states that objects attract one another with a force that varies as the product of their masses and inversely as the square of the distance between them.

Newton's law of universal gravitation may be expressed something like this:

All objects that have mass attract every other object with mass with a force that is directly proportional to the product of their masses and inversely proportional to the square of the distances between their centres.

Or mathematically:

$F = Gm_1m_2/r^2$

Where F equals the force; m_1 and m_2 are the masses of the objects; r is the distance between the objects; G is a number called the gravitational constant.

With three simple laws of motion and one law of gravity, Newton had found a way to explain the movement of everything in the universe. Newton's laws provided an explanation not only for the fall of an apple but also for the motion of the stars and planets.

LUNAR PROJECTILE

Galileo established that two forces govern the path of a projectile – gravity and the initial force that sends it on its way. The result of those two forces acting on it is that the projectile follows a curved path back to earth. Newton noticed that the distance the cannonball would fall in a given time was constant, since the acceleration due to gravity is constant, but the distance the projectile travels horizontally is dependent on its speed. Newton realized that at just the right velocity, the curved trajectory followed by the projectile would exactly match the curvature of the earth. The inertia of the projectile (which makes it continue travelling in a straight line) is balanced against the acceleration due to the earth's gravity (which pulls the projectile towards the centre of the earth). The projectile will now travel right around the earth, always accelerating towards the planet but never reaching the ground. The moon, Newton realized, is like a gigantic cannonball, perpetually falling around the earth in its orbit.

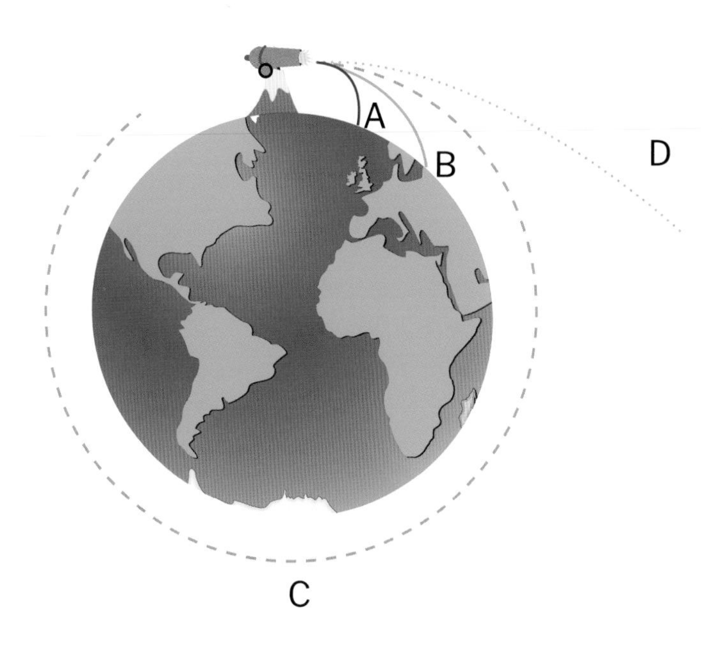

Newton calculated that a projectile fired at just the right velocity would follow a curved path that matched the curvature of the earth, effectively going into orbit. Too slow and it would fall to earth, too fast and it would escape into space.

Free falling

When the only force acting on an object is gravity, it is said to be in free fall. The orbiting moon is in free fall around the earth. A skydiver falling to earth is not quite in free fall since the frictional force of air resistance must also be taken into account.

In 1971, astronaut David Scott carried out a famous demonstration on the airless surface of the moon. He dropped a 1.32 kg (2.9 lb) hammer and 3 g (0.1 oz) feather and, with no air resistance to slow the feather, saw both hit the ground at the same time. It was confirmation that all objects accelerate at a uniform rate under gravity regardless of their mass.

Astronaut David Scott on the moon in July 1971.

5

Newton's bucket

Spinning in absolute space

Isaac Newton conceived of space as an absolute, unmoving reference point for the planets and other objects that exist within it. Every object, according to Newton, has an absolute state of motion relative to absolute space, so that an object must be either in a state of absolute rest, or moving at some absolute speed. Establishing the truth, or falsity, of this assertion would prove to be rather tricky.

In 1689, Newton described a seemingly simple experiment. Half fill a bucket with water and suspend it from a rope attached to a fixed point. Rotate the bucket, twisting the rope as much as possible, let the water settle, then let the bucket go. As the twisted rope unwinds the bucket rotates. At first, the bucket is

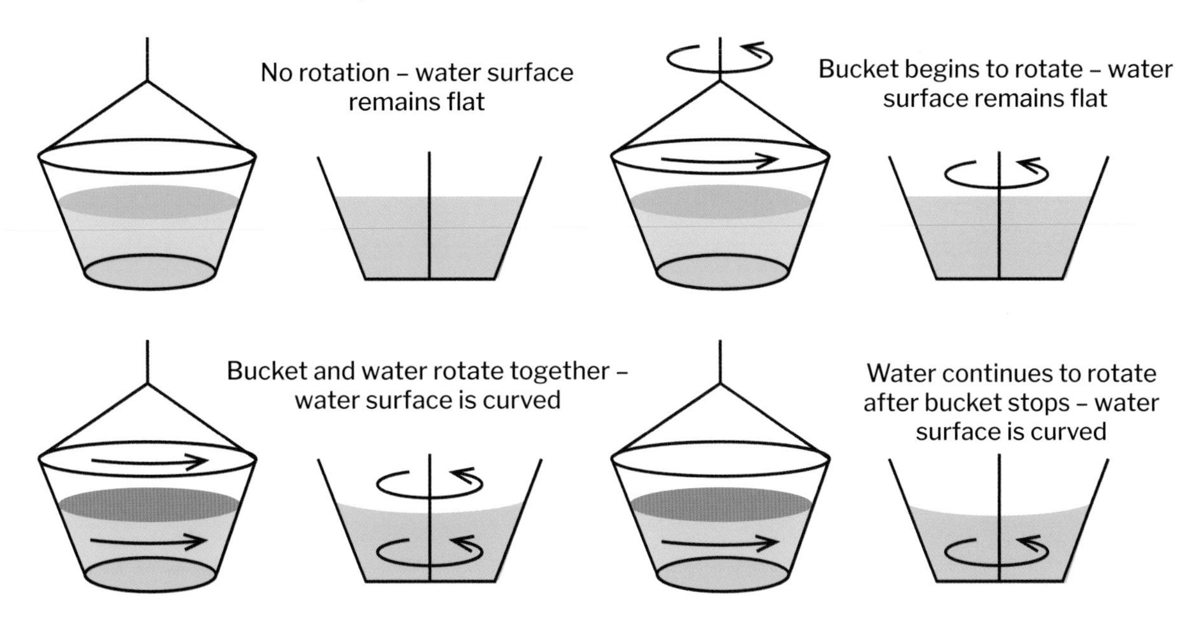

The surface of water in a spinning bucket gradually becomes concave as it spins. But why? Newton believed the water was spinning in relation to absolute space.

rotating relative to the stationary water and the surface of the water in the bucket remains flat, but slowly, as the water begins to rotate with the bucket, the water rises up the walls of the bucket and the surface becomes concave. After the bucket stops rotating the water goes on spinning relative to the bucket and the surface remains concave.

Newton posed a simple question: why does the surface of the water become concave? The obvious answer is that it is because the water is spinning. But what does this actually mean? When the bucket starts spinning the water is spinning relative to the bucket, yet to begin with its surface is flat. The shape of the water surface is not determined by the spin of the water relative to the bucket. The surface is concave when friction between the water and the sides of the bucket results in the two spinning together with no relative motion between them.

What would happen, Newton asked, if the experiment was carried out in empty space? He suggested a slightly different version in which two rocks are tied together with a rope, far from any gravitational influence. As the rope is rotated about its centre it becomes taut as the rocks pull outwards. But in the emptiness of space there is nothing to measure the rotation against, unless, as Newton concluded, it was measured against space itself. Newton used this as an argument for his belief in 'absolute space', claiming that what was meant by 'spin' was spin with respect to absolute space. When the water is not rotating with respect to absolute space its surface is flat, but when it spins with respect to absolute space its surface is concave. He wrote: 'Absolute space by its own nature, without reference to anything external, always remains similar and unmovable.'

Newton's notion of absolute space remained unchallenged for 200 years. The first serious challenger was Ernst Mach, who, in 1883, wrote that Newton's experiment with the rotating water bucket shows only that the rotation of water relative to the walls of the bucket produces no noticeable centrifugal forces; rather these forces were produced by its rotation relative to the mass of the earth and the other celestial bodies.

Mach believed that Newton was too ready to discard relative motion. It was not the rotation of the water relative to the bucket that should be considered but the rotation of the water relative to all the matter in the universe. If all that existed in the universe was the bucket and the water, the surface of the water would never become concave. If Newton's two rocks in space experiment were carried out in an empty universe, then, Mach concluded, it would be impossible to tell if the rocks were rotating. Rotation would be meaningless.

It was Albert Einstein's theory of special relativity in 1905 that finally laid the concept of absolute space to rest. However, as Newton's bucket involved

Isaac Newton in a portrait by Godfrey Kneller from 1702.

Ernst Mach.

the constant acceleration of circular motion, the special theory did not apply. As far as the spinning rocks were concerned, the special theory appeared to support Newton rather than Mach. The theory of special relativity replaced Newton's absolute space with a new notion of absolute spacetime. According to special relativity, observers moving at constant velocities relative to each other would not agree on the velocity of a bucket moving through space, but they would all agree on whether the bucket was accelerating or not.

Einstein's theory of general relativity, incorporating acceleration and gravity, was published in 1915. Einstein wrote to Mach in 1913, saying that Mach's view of Newton's bucket was correct and in line with general relativity. The behaviour of the spinning bucket is, as Mach claimed, determined by the gravitational forces of all the matter in the universe. General relativity does not, however, say that the two rocks in an empty universe experiment agrees with Mach. In this case it finds for Newton – the rope will become taut as the system spins. A universe with no matter has no gravity, in which case special relativity applies and all observers agree the rocks are spinning, or in other words accelerating. (The gravity produced by the mass of the rock system can be considered as negligible and will not produce the necessary forces to make the rope become taut.)

6

There and back again

Harmonic motion

Oscillations, movements back and forth in a regular rhythm, are happening all around us, all of the time. Harmonic, or periodic motion, a regular, repeated pattern of movement, occurs in many settings, both natural and artificial. A familiar example would be the vibrating strings of a musical instrument such as a guitar or a harp, or the simple back and forth movement of a pendulum.

Simple harmonic motion is an important type of periodic oscillation where the acceleration of the object in motion is proportional to its displacement from the equilibrium position. Timing the duration of one complete oscillation lets us determine the period and hence the frequency, the number of oscillations per

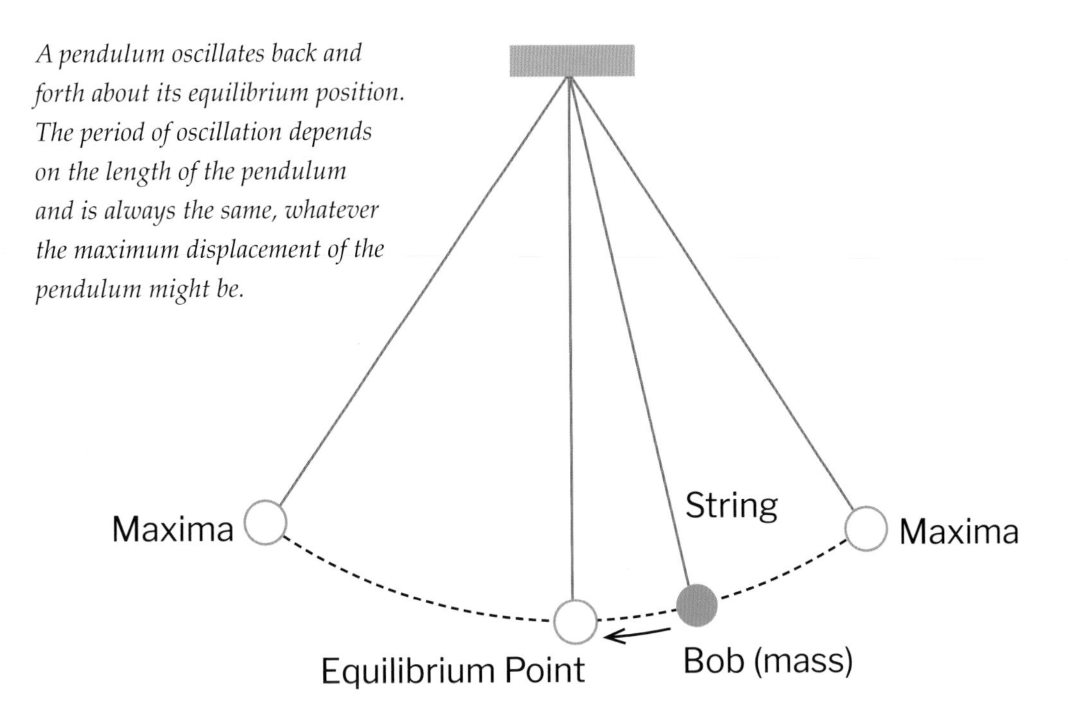

A pendulum oscillates back and forth about its equilibrium position. The period of oscillation depends on the length of the pendulum and is always the same, whatever the maximum displacement of the pendulum might be.

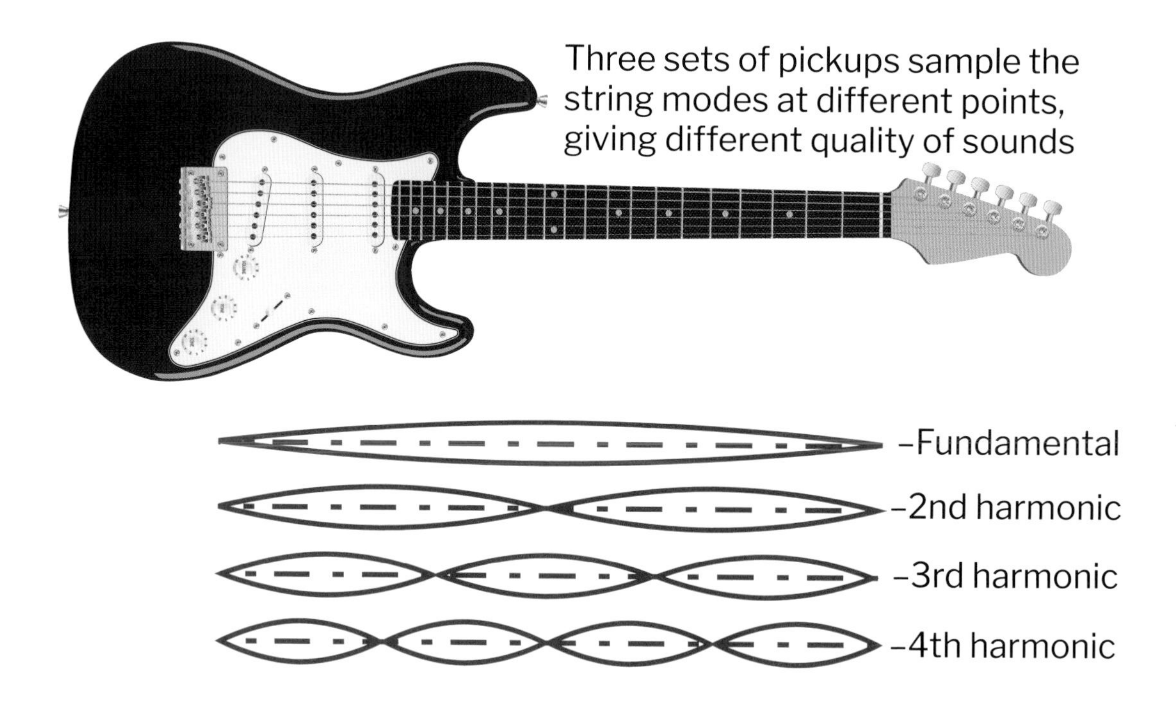

A plucked guitar string sets up a standing wave as the string oscillates around its equilibrium position. The string can be made to vibrate in different ratios to produce harmonics.

unit of time. The SI unit for frequency, the hertz (Hz), is defined as one cycle, or one complete oscillation per second.

In the case of the pendulum, the period is independent of the mass of the pendulum and is also independent of its amplitude, or maximum displacement from equilibrium. Each swing of a pendulum takes exactly the same length of time (this is why pendulums were used as timekeepers in mechanical clocks for centuries). Plotting the angle of the pendulum's swing against time on a graph produces a pattern known as a sine wave.

STANDING WAVES

In 1732, Swiss physicist Daniel Bernoulli found a way of applying Newton's second law of motion (see page 15) to explain how the strings of a musical instrument produce different sounds as they vibrate at different frequencies. Bernoulli demonstrated how the force acting on the string grew as it moved further from its stationary starting point and always acted in the opposite direction to the string's displacement from the central equilibrium position. The force tended to restore the string to the equilibrium position but as the string overshot a repeating cycle was set up, producing a simple harmonic motion, just like that of a swinging pendulum though with much greater frequency.

Plucking a guitar string sets up a standing wave, in which some points remain fixed while others between them vibrate with the maximum amplitude. These fixed points are called nodes and the points of maximum amplitude are called antinodes. It's important not to confuse nodes and antinodes with crests and troughs. A standing wave is not actually a wave. It is an interference phenomenon, a pattern formed by two or more waves of the same frequency but with different directions of travel within the same medium.

An antinode vibrates back and forth at the same point on the string, it does not travel along it as a wave would.

Daniel Bernoulli.

Energy transfers

Plotting the changes in potential energy and kinetic energy of a swinging pendulum on a graph produces sine waves. At maximum displacement, all of the energy in the system is in the form of potential energy and the velocity of the pendulum is zero. As the pendulum swings back towards the equilibrium position, where the pendulum reaches its maximum velocity, the potential energy is converted into kinetic energy. Adding the potential energy and kinetic energy curves together produces a straight line – an elegant illustration of the conservation of energy (see page 32), showing how one form is transferred to the other.

7

Neither created nor destroyed

The nature of energy

What is energy? On the face of it, it seems like a simple enough question, but the more you think about it the more perplexing it becomes. It is often defined as 'the ability to do work', but what is it that delivers that ability? And what do we mean by 'work'? All we can really measure are the effects energy has – but not the thing itself. We have no knowledge of what energy is, only what it does.

There is no such thing as 'pure energy' existing on its own, nothing that can be put in a bottle and measured. Though energy itself is an enigma, we can define the specific forms it takes:

Mechanical energy is the energy of mechanical systems, such as a bullet fired from a rifle, or a tennis ball struck by a racket. Mechanical energy takes three forms:

- **Gravitational potential energy**: the energy stored in an object or system due to gravity. A diver on the high board stores gravitational potential energy. When they launch towards the water this is converted into:

- **Kinetic energy:** the energy due to the motion of an object. The faster an object moves, or the greater its mass, the more kinetic energy it possesses. Electrical energy can be thought of as a type of kinetic energy resulting from the movement of electrical charges.

- **Elastic potential energy**: the energy stored in a stretched spring, rubber band, or other elastic material.

- **Thermal energy**: the energy resulting from the kinetic energy of molecules of a substance. Hot water has more thermal energy than cold water as the molecules are moving faster and have greater kinetic energy.

- **Radiant energy**: the energy of electromagnetic radiation, such as visible light, microwaves or x-rays.

- **Chemical energy**: the energy stored in chemical bonds. The food we eat is a store of chemical potential energy released in respiration.

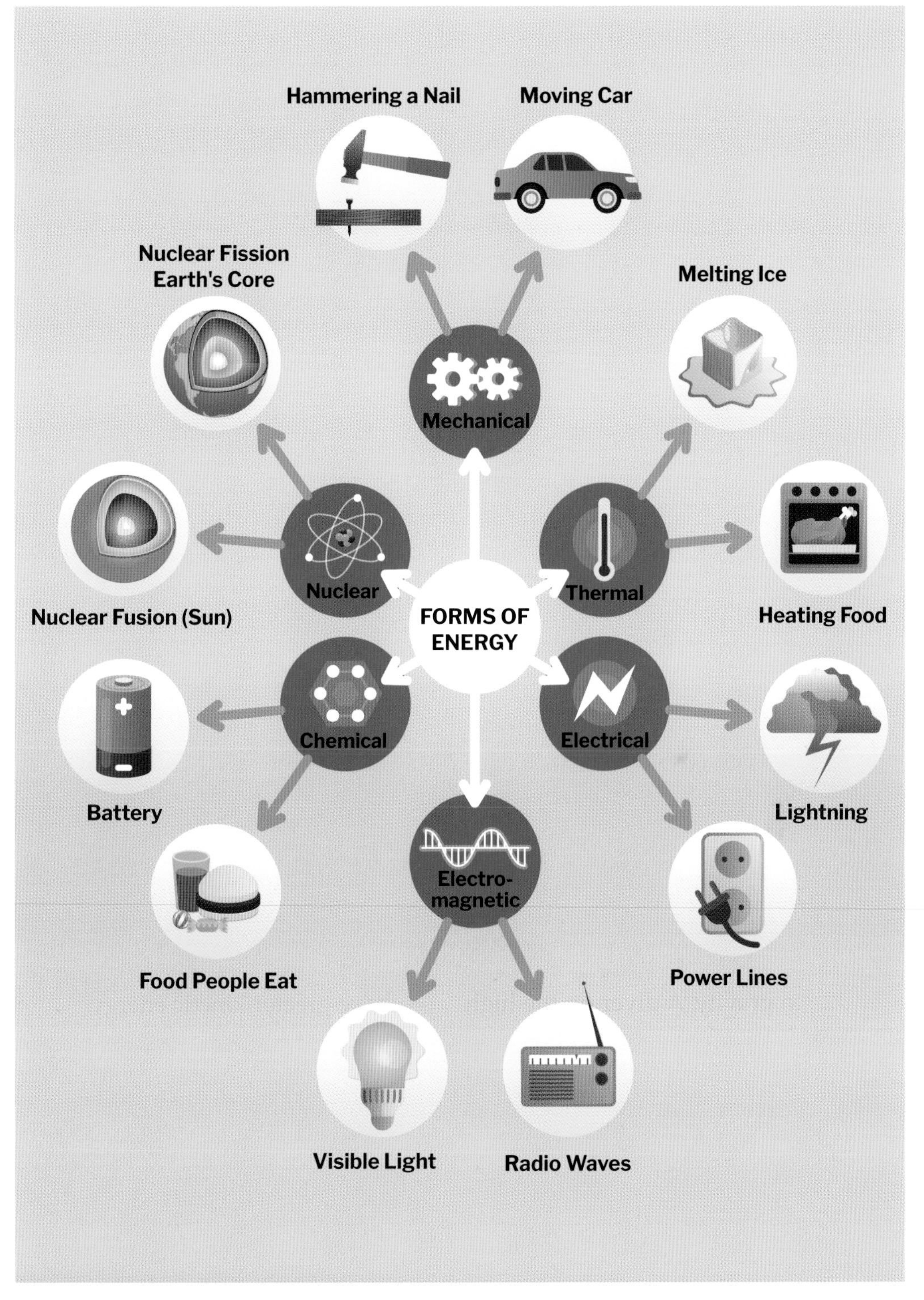

The many different types of energy.

> *'There is… a law, governing all natural phenomena... There is no known exception to this law… called the conservation of energy. It states that there is a certain quantity, which we call energy, that does not change... it is not a description of a mechanism, or anything concrete; it is just a strange fact that we can calculate some number and when we finish watching nature go through her tricks and calculate the number again, it is the same.'*
>
> Richard Feynman

- **Nuclear energy**: the energy stored inside atoms that is released during nuclear reactions. It powers nuclear reactors, atomic bombs and the stars.

CONSERVATION OF ENERGY

Energy can be transferred from one form, or energy store, to another – electrical energy can be transferred into light energy; chemical energy can be transferred into kinetic energy, for example. It is this transfer of energy, moving from one form to another, that makes things happen. But whatever happens, the total amount of energy in a closed system does not change, no matter what transfers take place. This is known as the law of the conservation of energy, and it is one of the unalterable laws of physics.

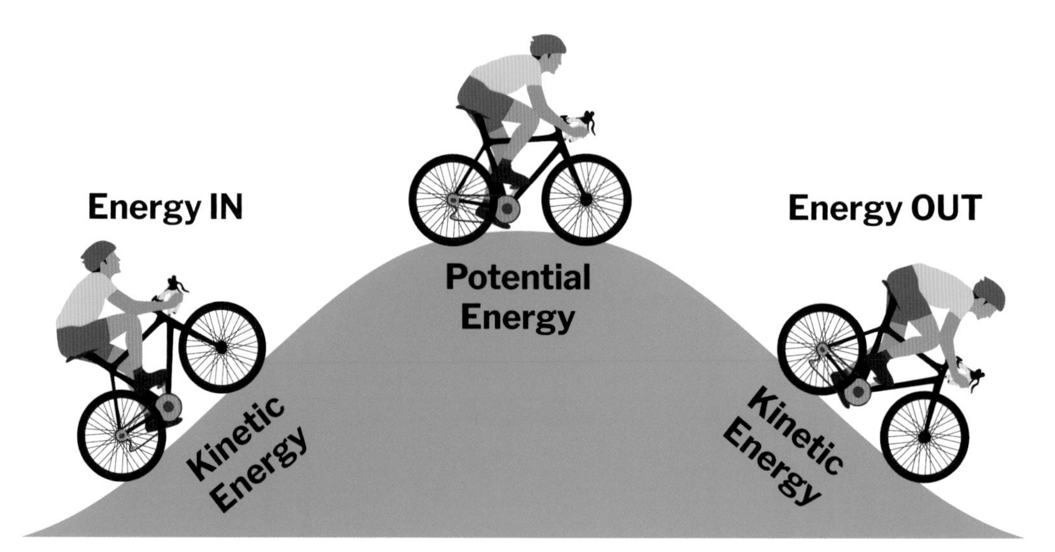

Kinetic and potential energy.

James Joule.

GETTING WORK DONE

Energy gets a rocket to the moon, a bicycle to the shops, or a lift to the top of a building. The energy transferred when a force moves an object is called work. Work is a measure of energy, expressed in units called joules, after James Joule. The total energy transferred when work is done is given by the formula:

$$W = F \times d$$

W = work (in joules); F = force (in newtons) and d = distance (in metres)

How quickly the work is done, i.e. how quickly energy is transferred from one form to another, is measured as power. The more energy transferred in unit time the greater the power. Power is measured in watts – 1 watt equals a transfer of 1 joule per second.

$$P = E/t$$

Power (in watts) = Energy transferred (in joules)/time (in seconds)

The kilowatt-hour familiar from energy bills means using energy at the rate of 1 kilowatt (1,000 watts, or 1,000 joules per second) for a period of 1 hour which equals 3,600,000 joules.

8

Going with the flow

Fluid mechanics

For a physicist, a fluid is a material that flows. The particles of a fluid are in continual motion. Unlike the particles that make up a solid, the particles of a fluid are not bound together, and it can change shape readily when subjected to a force.

Fluid mechanics is the study of the behaviour of fluids and can be divided into two branches: fluid statics, which deals with fluids at rest, and fluid dynamics which is concerned with fluids in motion. Fluid mechanics has applications across a wide range of fields, including biological systems,

The first Eureka moment

The earliest known work on the physics of fluids was *On Floating Bodies*, by the Greek philosopher Archimedes. Here he states his finding that a body immersed in a liquid displaces a volume of water equal to its own volume and feels an upward buoyant force equal to the weight of water it displaces, a discovery that Archimedes legendarily made while in his bath, occasioning his cry of 'Eureka!' (I have found it!)

The buoyant force felt by an object immersed in a fluid is equal to the weight of the displaced fluid. If the weight of the object is less than that of the displaced fluid it will float.

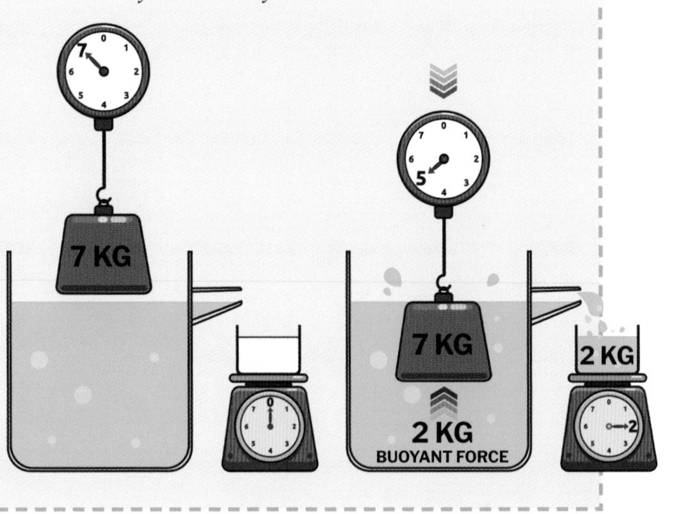

meteorology, mechanical and chemical engineering, and astrophysics.

A fluid has no fixed shape and will take up the shape of its container, yields easily to external pressure, and flows from one point to another. The most common types of fluid are liquids and gases. One of the most definitive characteristics of a fluid is that it flows.

The movement of liquids and gases, generally referred to as 'flow', describes how fluids behave and how they interact with their surroundings, for example, the movement of gas through a pipe, or water spilling over a surface. Flow can be either steady or unsteady. Steady flows do not change over time. An example of steady flow would be gas flowing through a pipe at a constant rate. An example of an unsteady flow would be someone using a hand pump to reinflate a bicycle tyre.

TORRICELLI'S LAW

Imagine a container of water that has holes punctured at different heights.

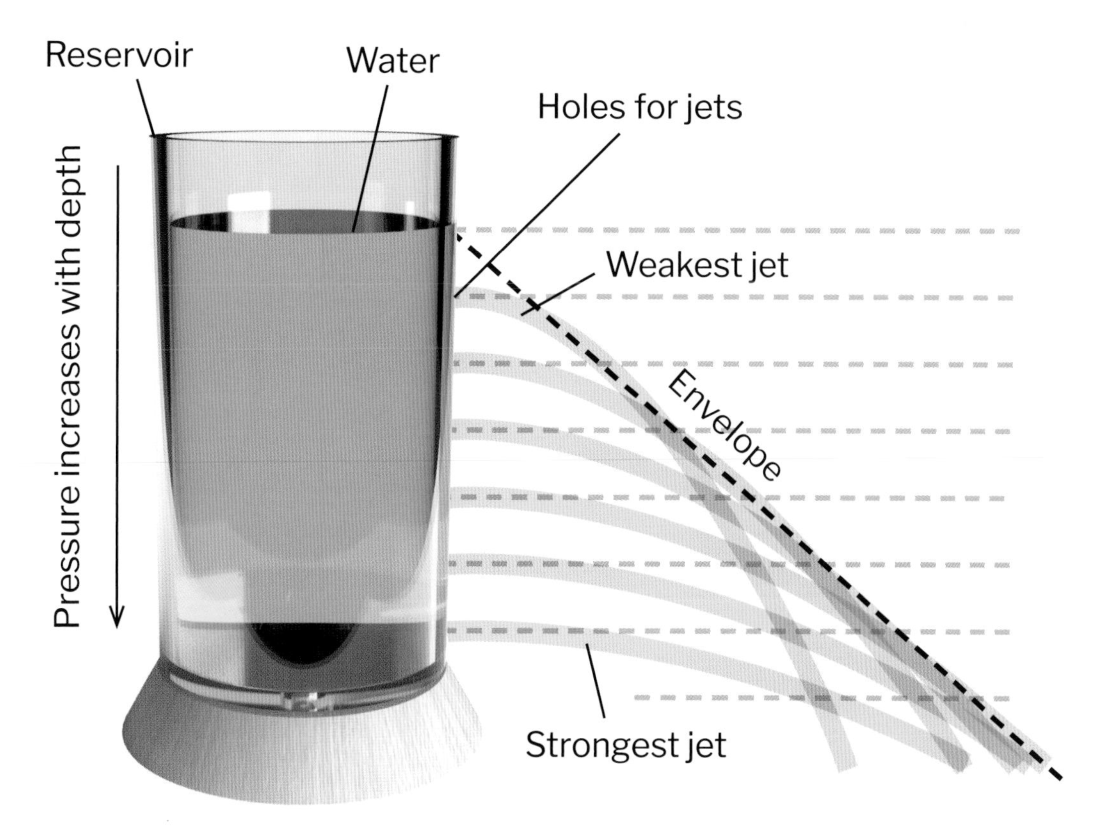

Torricelli's law established that a fluid will escape more forcefully from a hole lower down a container than from one nearer the top. All of the jets fall within an envelope, a line drawn at 45 degrees from the water surface.

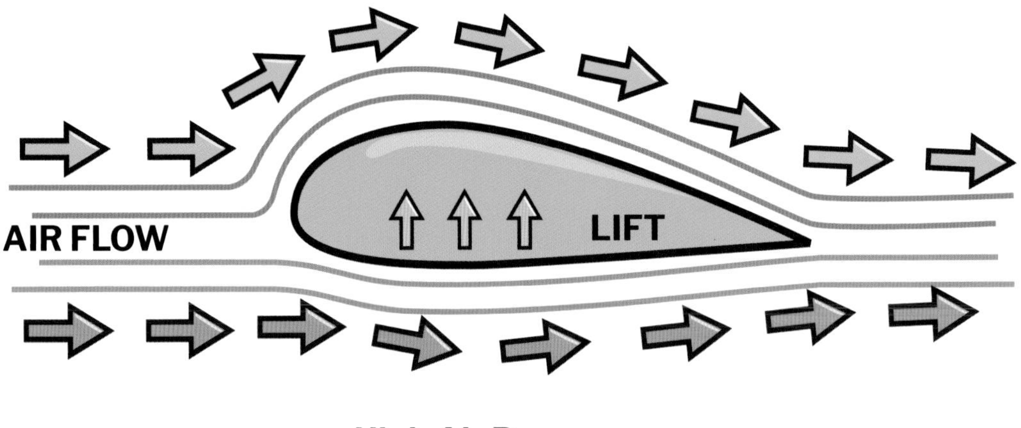

The faster-flowing air over the top of an aircraft's wing exerts less pressure than the slower-moving air beneath, resulting in a lifting force.

Where does the water come out fastest? Torricelli's law, formulated by Italian mathematician Evangelista Torricelli in the mid-17th century, established the velocity of a fluid leaving through a hole in the container is the same as it would be if dropped from the same height to the hole level. So, the lower down the container the hole is, the faster the stream of fluid. As an equation, Torricelli's law can be expressed as: $v = \sqrt{(2gh)}$ (where v is the velocity of the fluid, g is the acceleration due to gravity, and h is the height of the fluid above the hole).

BERNOULLI TAKES FLIGHT

Swiss physicist Daniel Bernoulli laid much of the groundwork for our understanding of fluids in his 1738 work, *Hydrodynamics*. It was here that he set out what became known as Bernoulli's principle which says that an increase in the speed of flow of a fluid results in a reduction in its pressure and a decrease in the speed of flow results in an increase in pressure.

It is the Bernoulli principle that makes it possible for flight. For an aircraft to take to the air, a force must be generated that exceeds the force of gravity. This force is called lift. The shape of an aircraft's wing causes air to flow faster across the top than it does over the bottom. The faster-flowing air results in a reduction in the air pressure over the wing and because the air pressure below the wing is greater than above it, a resulting lift force is generated.

9

Getting warmer

Thermal radiation

Heat energy is transferred from one place to another in three ways: by conduction in solids, by convection in fluids, and by thermal, or heat, radiation. Thermal radiation does not require physical contact. Like radio waves, visible light and x-rays, thermal radiation is a form of electromagnetic radiation and travels in waves through space.

The amount of thermal radiation given off by an object depends on its temperature. The hotter the object, the more energy it emits. If an object is hot enough the radiation it emits can be seen as visible light. A metal rod heated to a high enough temperature will begin to glow, first a dull red, then yellow and

Thermal scanners were used at airports to screen passengers who may have had Covid-19.

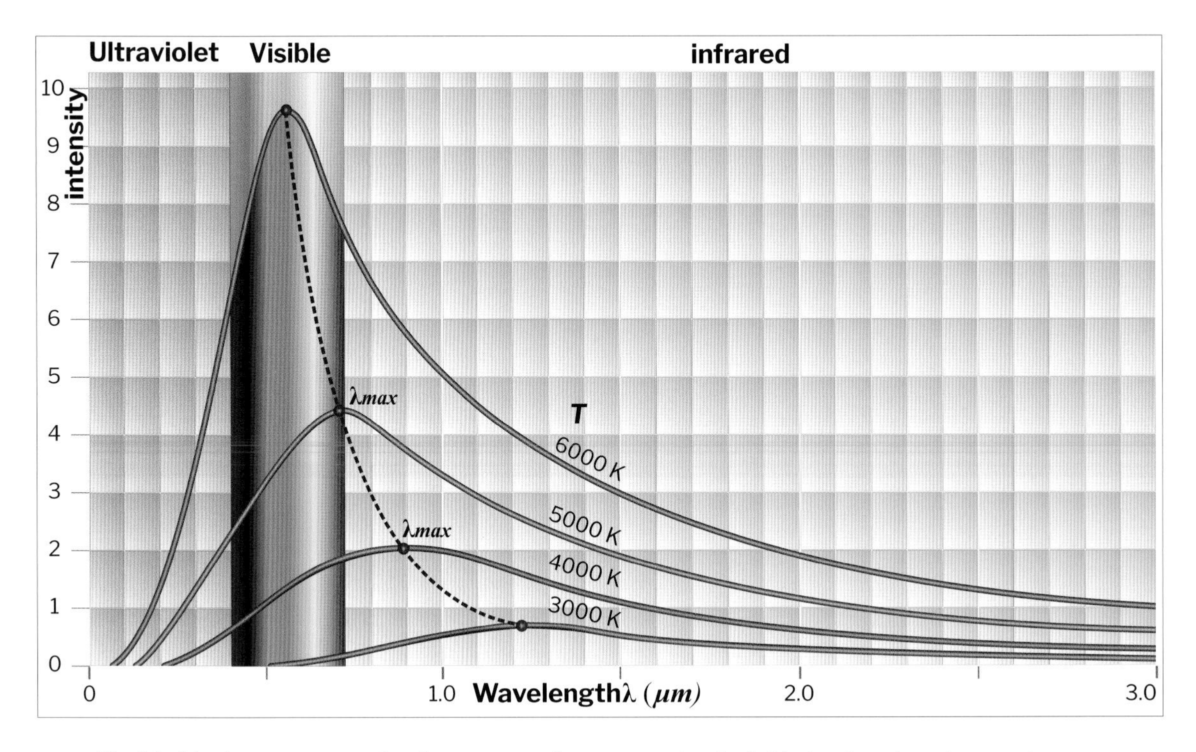

The blackbody curve maps the electromagnetic energy output of objects at various temperatures.

finally a brilliant white.

All of the objects in the universe are exchanging electromagnetic radiation with each other all the time. This constant flow of energy from one object to another prevents anything from ever cooling to absolute zero (equal to -273.15°C / -459.67°F), the theoretical minimum of temperature at which an object transmits no energy at all. Everything that has a temperature above absolute zero emits radiation.

BLACK BODIES

A material that tends to absorb energy at a certain wavelength tends also to emit energy at that same wavelength. It is possible to imagine an object that perfectly absorbs all of the electromagnetic radiation that strikes it, with none being reflected back. If no radiation is reflected from it, all the energy that it emits depends solely on its temperature. Physicists call these hypothetical objects black bodies. An ideal black body absorbs and emits energy with 100 per cent efficiency. No such object exists in nature. Most of the energy output of a black body is concentrated around a peak frequency, which increases as the temperature increases. The spread of energy emitting wavelengths around the peak frequency form a distinctive shape called a blackbody curve. The blackbody curve of the sun, for example, peaks at the centre of the visible light range.

The law of thermal radiation, first stated by German physicist Gustav Kirchhoff in 1860, says that the efficiency with which an object absorbs radiation at a given wavelength is the same as the efficiency with which it emits energy at that same wavelength. This means that for an object in thermodynamic equilibrium, that is at the same temperature as its surroundings, for any given temperature and wavelength, the amount of radiation absorbed by the surface is equal to the amount emitted.

AVERTING CATASTROPHE

In 1893, German physicist Wilhelm Wien worked out the mathematical relationship between temperature change and the shape of the blackbody curve. It allowed the peak wavelength to be calculated for any temperature, and it explained why things change colour as they get hotter. As the temperature increases the peak wavelength decreases, going from longer infrared waves to shorter blue-white and ultraviolet. Within a few years, however, careful experiments were showing that Wien's mathematical predictions weren't holding true for the infrared range. In 1900, British physicists Lord Rayleigh and James Jeans came up with a revised formula that seemed to explain what was going on at the red end of the spectrum, but they created problems of their own. If Rayleigh and Jeans' theory was correct, there was effectively no upper limit to the higher frequencies generated by the blackbody radiation. If this were so, then opening the oven to check on your cakes would result in instant annihilation in a burst of an infinite number of highly energetic waves. This came to be known as the ultraviolet catastrophe, and it was obviously wrong.

That same year, Max Planck in Berlin was also working on a theory of black body radiation. He came up with an explanation for the black body curve that agreed with all the experimental measurements, and at the same time changed physics for ever. Planck found that catastrophe could be averted by assuming that a black body emitted energy not in continuous waves, but in discrete packets, which he called quanta. The day he presented his findings to a meeting of the Deutsche Physikalische Gesellschaft (the German Physical Society) in Berlin – 14 December 1900 – marked the birth of quantum mechanics and the beginning of a new era in physics.

Heat transfers

Heat always transfers from hot objects to cold objects until they reach thermal equilibrium. Heat energy can be transferred in one of three ways:

Conduction – the transfer of thermal energy by physical contact. Dense crystalline solids, such as metals, are good thermal conductors. When a metal bar is heated at one end the energy causes the particles that make it up to vibrate faster, causing neighbouring particles to vibrate too. This results in the transfer of energy along the metal bar. A material that conducts poorly, such as air where the particles are far apart, is called an insulator.

Convection – the transfer of thermal energy via moving currents in a fluid. When part of a liquid or gas is heated it becomes less dense than the material surrounding it, causing it to rise and create a convection current. Thermals are rising columns of warm air produced by convection as a warm surface heats the air above it.

Radiation – the transfer of thermal energy by infrared radiation. All objects emit infrared radiation; the hotter an object is the more radiation it gives out. When an object absorbs infrared radiation its temperature rises.

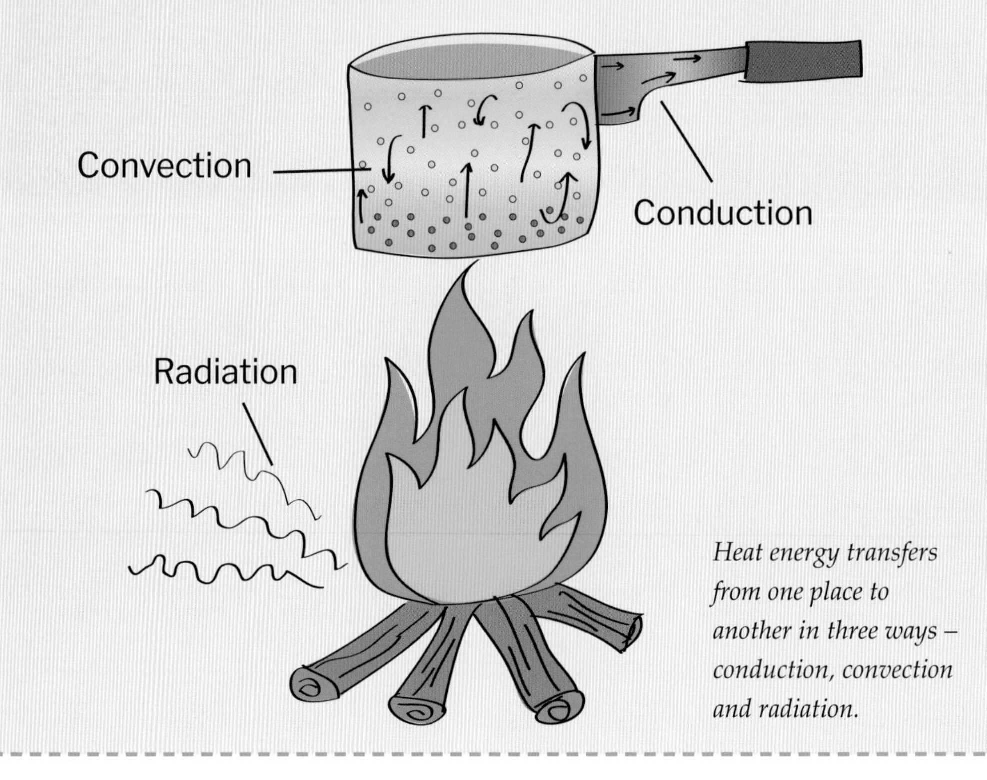

Heat energy transfers from one place to another in three ways – conduction, convection and radiation.

10

Heat and work

The first law of thermodynamics

The 18th-century invention of the steam engine, powerhouse of the Industrial Revolution, spurred the development of the science of thermodynamics. The nature of heat, energy and work was investigated more thoroughly than ever before, as scientists and engineers looked for ways to squeeze every last drop of efficiency from their new machines.

Before the Industrial Revolution, mechanical power was supplied by natural energy sources such as wind and water or the muscles of people and animals. The invention of the steam engine changed that. The trouble was that steam engines were hugely inefficient; only around 3 per cent of the fuel burned was converted into useful work. Various ways were tried to improve the efficiency of the engines, but they were hampered by the fact that there was little insight into the key energy process involved – the transfer of heat. It was the need to acquire this knowledge that resulted in the development of thermodynamics – the science of heat.

The foundations of thermodynamics were laid in 1820 by a young French soldier called Nicolas Léonard Sadi Carnot. Carnot wanted to find a way of improving the efficiency of French steam engines, which he thought suffered in comparison to their British counterparts, and directed his attention to the movement of heat through the engine. He saw parallels with the movement of water which always flows downhill and can be used to do work. Carnot imagined 'calorific fluid', as heat was then thought to be, flowing through the steam engine from hot to cold areas in a similar fashion.

Carnot imagined an idealized heat engine from which all friction had been eliminated so that the work output was the same as the heat input with no energy lost in the conversion. Carnot realized that it was impossible to prevent some heat being lost to the environment, but his ideas helped designers improve the efficiency of their engines.

THE FIRST LAW OF THERMODYNAMICS

The idea that energy is conserved was first formulated in 1841 by German

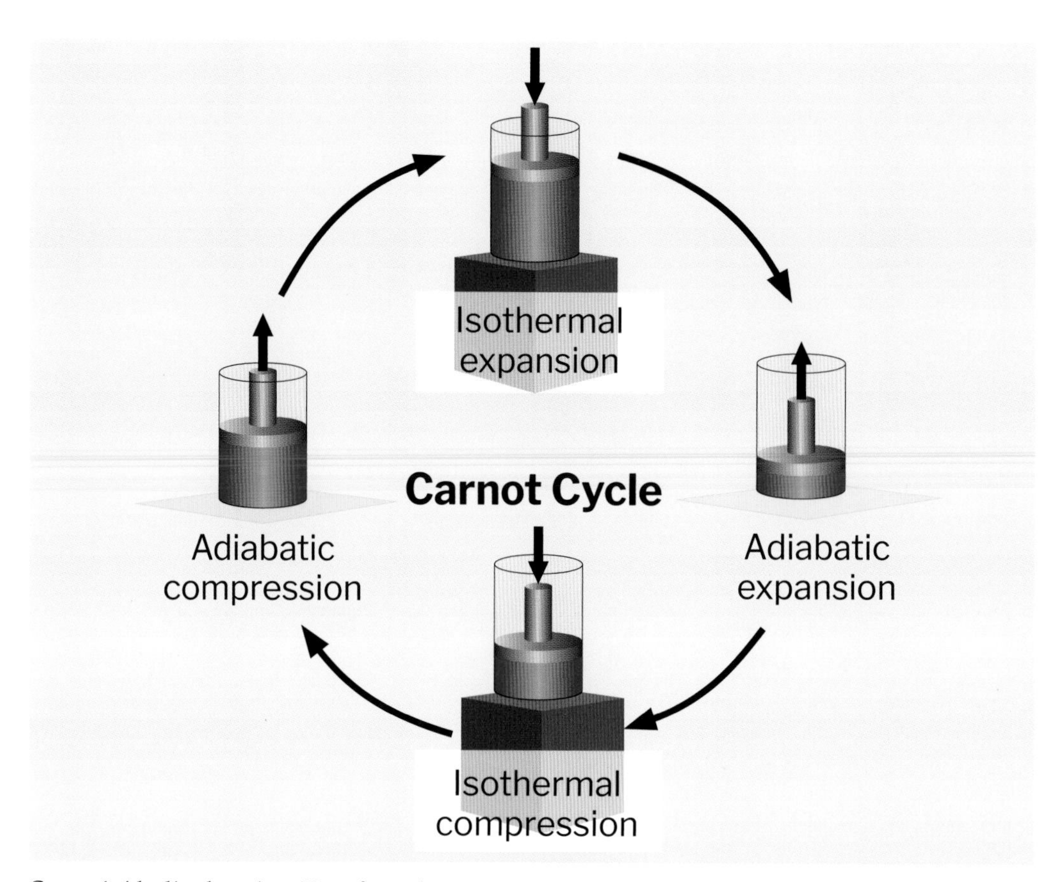

Carnot's idealized engine. Heat from the source causes the gas to expand, pushing the piston out (isothermal expansion). Without further heat input the gas continues to expand, cooling as it does so (adiabatic expansion). Reversing the process and compressing the gas causes a loss of heat (isothermal compression). When the system is isolated from further heat loss the temperature rises back to its original level (adiabatic compression).

physician Julius Robert von Mayer. He considered that the food we eat produced both body heat and gave us the ability to do work. He came to the view that heat and work are interchangeable – that food can produce either heat or work, but the total energy available has to remain the same. He believed this principle applied not just to living things but to all systems using energy and set out what he called a 'conservation law of force' (by which he meant energy).

At around the same time in England, brewer's son James Joule was the first to back up von Mayer's ideas with firm experimental evidence, showing that heat and mechanical work were equivalent: a given amount of work could be transformed into a measurable and predictable amount of heat. It was the work of researchers such as von Mayer and Joule that paved the way towards the conservation of energy law, which in its most direct form can be stated as:

Energy can be neither created nor destroyed.

The first law of thermodynamics applies the conservation of energy law to thermodynamic processes. As an equation the first law can be stated as:

$$\Delta U = Q - W$$

The change in internal energy of a system (ΔU) is equal to the heat added to the system (Q) minus the work done by the system (W). The internal energy of a system is the energy associated with the random motion of the molecules that it consists of. It is distinct from the kinetic or potential energy of the system as a whole.

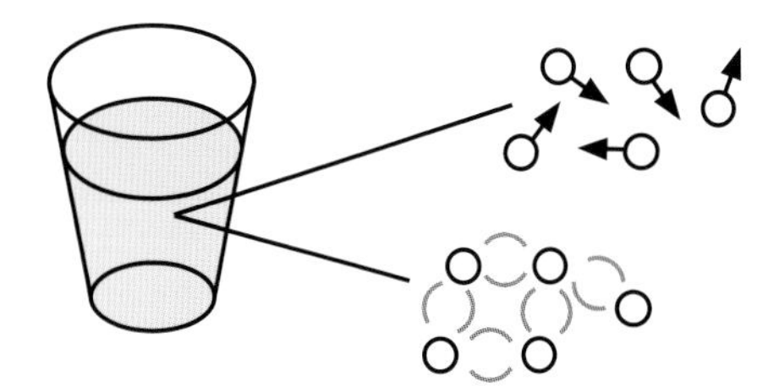

The internal energy of a glass of water, comprised of the kinetic energy of the molecules and the attractive forces between them, is distinct from the kinetic energy of the whole glass if it is knocked to the floor.

The zeroth law

The availability of reliable thermometers in the 18th century allowed for more precise measurements of temperature and heat flow. This led to observations such as those of Joseph Black at the University of Edinburgh, who noticed that a collection of objects at different temperatures, if brought together, will all eventually reach the same temperature, or thermal equilibrium, at which point no further heat flow takes place. This eventually became the zeroth law of thermodynamics which states that if two objects are in thermal equilibrium with a third then they are in thermal equilibrium with each other.

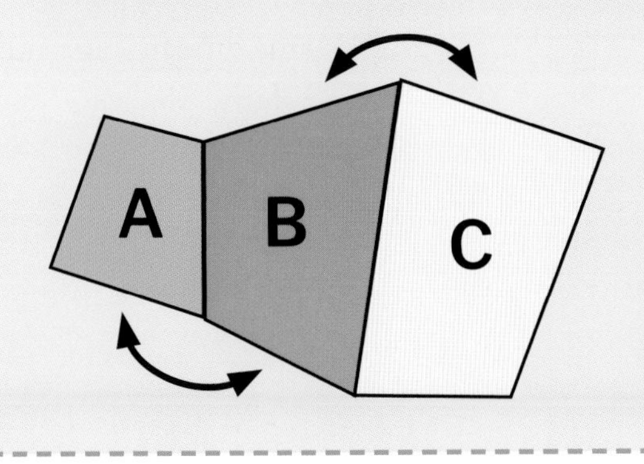

If A is in thermal equilibrium with B and B is in thermal equilibrium with C, then it follows the C will also be in thermal equilibrium with A.

11

Spring in the air

The gas laws

The insubstantial nature of gases may explain why it took so long for researchers to begin to investigate their physical properties. Although air had its place as one of the four elements of Classical Greece (alongside, earth, fire and water) no real study of the nature of gases was made until the 17th century.

Like solids and liquids, gases have physical properties. A gas is made up of particles, unattached in any way to each other, in constant random motion,

The barometer

The barometer was one of the crucial inventions that made research into gases possible. It is generally credited to Italian physicist and student of Galileo Evangelista Torricelli around 1644. Torricelli filled a glass tube with mercury and inverted it in a small basin which was also filled with mercury. He observed that the mercury in the tube maintained a height of about 76 cm (30 in) which he correctly deduced was due to the pressure of the atmosphere pushing down on the mercury in the basin. Later experiments showed that the level of the mercury column varied with changes in atmospheric pressure at different altitudes and in different weather conditions. Torricelli's invention was soon seeing service in the laboratory, as an instrument for measuring altitude, and for monitoring and predicting the weather.

Evangelista Torricelli, inventor of the barometer.

colliding with each other and with the walls of any container in which they are placed. It is these collisions that create pressure. During the 17th and 18th centuries, through painstaking experiment, scientists Robert Boyle in Britain, and Jacques Charles and Joseph Gay-Lussac in France set out the gas laws that determine how gases behave.

BOYLE'S LAW

The discovery of the first of the gas laws by Robert Boyle in 1662 came about thanks to the invention of the barometer. Boyle thought that air was made up of coiled particles that could be compressed, but which would spring back when the pressure was released. 'There is a spring… in the air we live in', he wrote. Boyle's experiments established a simple relationship in which the volume of a gas and its pressure vary in inverse proportion. In other words, if the pressure is doubled, at constant temperature, the volume of the gas is reduced by a half; if the pressure is tripled, the volume is reduced to a third, and so on. As an equation this is expressed as:

$$pv = k$$

(pressure x volume = k, a constant)
This is the equation that became known as Boyle's law.

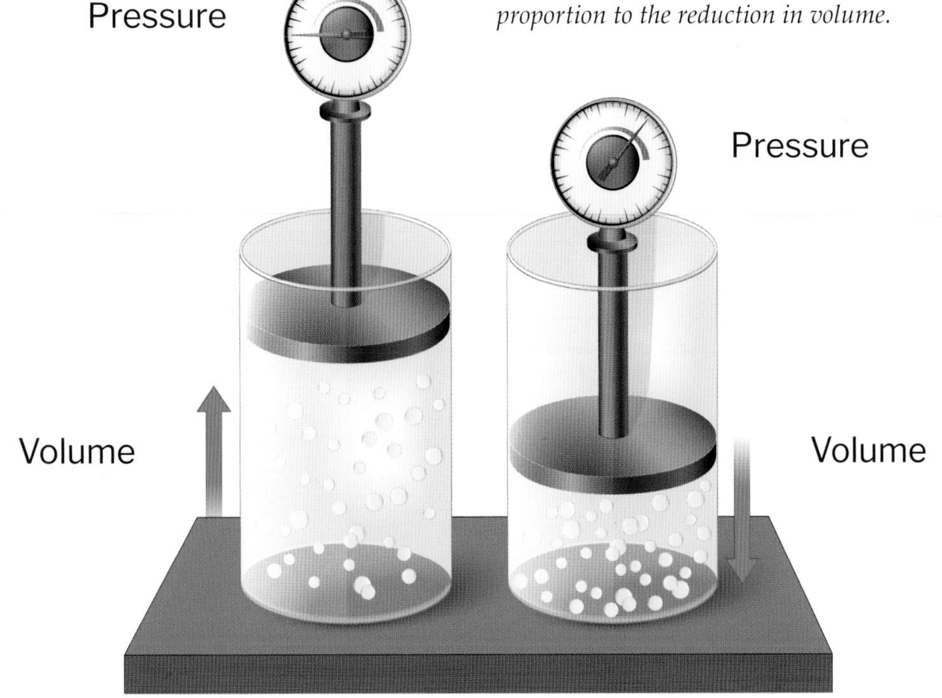

Boyle's law states that if a gas is compressed while its temperature is held constant the pressure it exerts will increase in inverse proportion to the reduction in volume.

Keen balloonist and physicist Jacques Alexandre César Charles observed that the volume of a gas is directly proportional to its temperature. A hot-air balloon rises because the air inside it is less dense that the surrounding atmosphere.

CHARLES' LAW

In the early 1700s French physicist Guillaume Amontons developed the air thermometer. This measured temperature based on a proportional change in pressure, a relationship known as Amontons' law: $p/t = $ a constant. According to Amontons' law, increasing the temperature of a fixed volume of gas will increase its pressure. Amontons' law explains why you should only check your tyre pressure when the tyres are cold. The friction of the tyres on the road surface increases the temperature of the air inside the tyre, resulting in the air pressure in the tyre also increasing. Tyre pressures set when the tyres are warm are likely to be too low.

In 1783, French physicist and

inventor Jacques Alexandre César Charles launched the first manned hydrogen balloon from the Jardin de Tuileries in Paris. Charles and his fellow aeronaut Nicolas-Louis Robert took a barometer and a thermometer aloft with them which they used to measure the pressure and the temperature of the air at altitude. This was not only the first flight of a manned hydrogen balloon but also the first flight of a weather balloon.

Inspired by his ballooning exploits, Charles carried out experiments in the course of which he observed that the volume of a gas is directly proportional to its temperature, a finding that became known as Charles' law:

$$v/t = a\ constant$$

In principle, this was basically the same finding that Amontons had made some years earlier. This relationship between the temperature and the volume of a gas explains how hot-air balloons work. As discovered by Archimedes, an object that weighs less than the fluid it displaces will float. Because a gas expands when heated, a given mass of hot air occupies a larger volume than the same mass of air when cold. So the hot air inside a balloon is less dense than the cold air that surrounds it. Thus, the balloon begins to rise, floating on the cooler, denser air around it, like a cork bobbing on a pond.

THE PRESSURE LAW

Joseph Louis Gay-Lussac, who was also a keen balloonist, set out what is known as the pressure law, which states that if the mass and volume of a gas are kept constant the pressure of the gas will vary according to its temperature. Because Amontons had established the general principle of the pressure law a century earlier it is perhaps more properly known as Amontons' law. What is generally referred to as Gay-Lussac's law more usually refers to what is also known as the law of combining volumes, which he formulated in 1808. This states that, when gases react together to form other gases, and all volumes are measured at the same temperature and pressure, the ratio between the volumes of the reactant gases and the products can be expressed in simple whole numbers. For instance, two volumes of hydrogen combine with one volume of oxygen to create two volumes of water vapour in a simple 2:1 ratio.

Italian scientist Amedeo Avogadro explained Gay-Lussac's finding in 1812 when he theorized that at a given temperature and pressure, equal volumes of all gases have the same number of molecules and that the number of molecules present varies exactly with the volume. According to Avogadro's hypothesis, oxygen existed as a double-atom molecule that split to combine with two atoms of hydrogen to produce one molecule of water. This had to be the case if there were as many molecules of water as there were of oxygen and hydrogen.

An ideal gas

It is obvious that the gas laws are linked. Boyle's law linked pressure and volume, Charles' law linked volume and temperature, and Amontons' law linked temperature and pressure. The three can be written together as the combined gas law as pv/t = a constant.

A gas that obeys the combined gas law is referred to as an ideal gas. An ideal gas has a number of properties:

1. An ideal gas consists of a large number of identical molecules.
2. The individual molecules themselves take up no volume.
3. The molecules obey Newton's laws of motion, and they move randomly.
4. The molecules do not attract or repel each other; any collisions are completely elastic, and take a negligible amount of time.

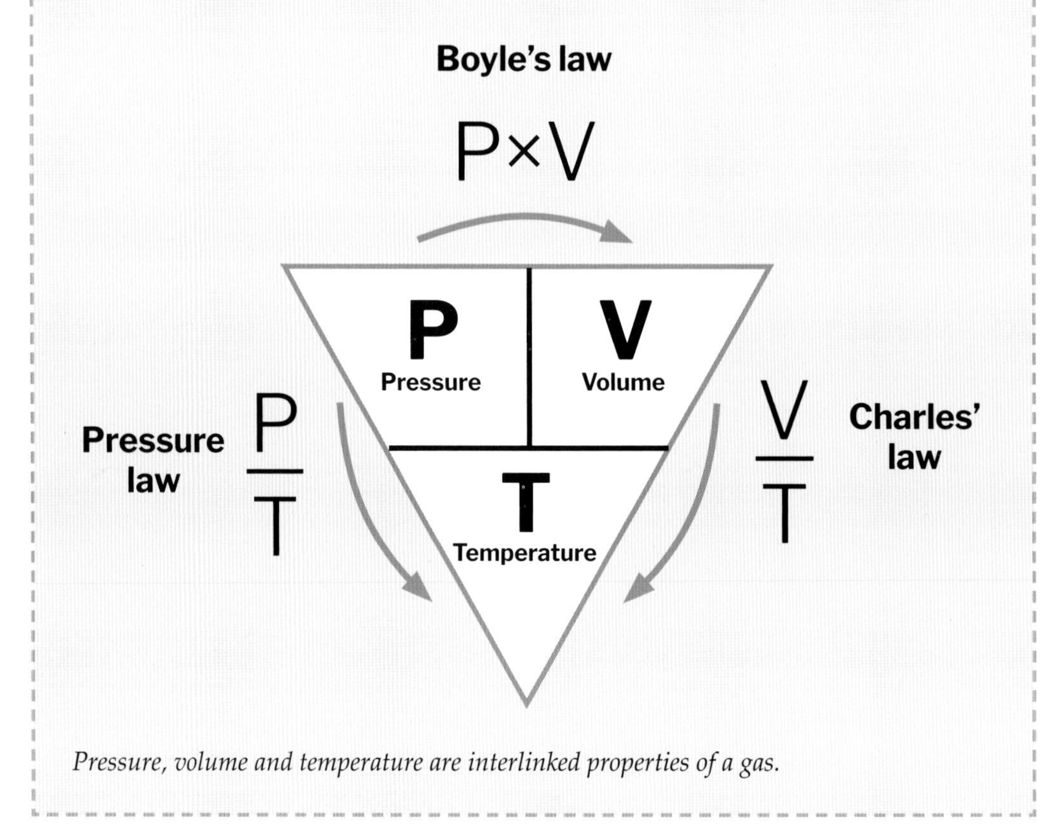

Pressure, volume and temperature are interlinked properties of a gas.

12

The end of all things

Time's arrow and the second law of thermodynamics

Some of the laws of physics, such as Newton's laws of motion, work just as well backwards as forwards in time. If we know how an object is moving now, we can map its trajectory back into the past, and forward into the future. Time itself has no direction in Newton's universe. So why does time flow only in one direction?

THE SECOND LAW OF THERMODYNAMICS

German physicist Rudolf Clausius was one of the founding thinkers of thermodynamics. In 1850, he published a paper setting out his ideas on heat and work, including the shortcomings he saw between Carnot's ideas and the conservation of energy law. The energy conservation law would work equally well if heat flowed from a low temperature area to a high one, which, in reality, was never seen to happen. Clausius therefore proposed a second law of thermodynamics which stated that:

Heat always flows from a hot object to a cooler one and never the other way around.

Eventually the system reaches a state of equilibrium. Clausius also theorized that heat was a result of the kinetic energy of the moving particles that made up a substance, which turned out to be the case.

The second law dictates that most natural processes are irreversible as to reverse them would involve the reversal of an energy flow, which the second law forbids. Entropy (broadly defined as a measure of disorder) and energy are similar in that an object or system may be said to have a certain 'entropy content' just as it has a certain 'energy content'. However, while the first law of thermodynamics ensures that the energy of a system is always conserved, the second law of thermodynamics ensures that the total entropy of an isolated system cannot decrease: it may (and generally does) increase.

THE REVERSIBILITY PARADOX

Just as with any other objects in motion, collisions between individual molecules are completely reversible. In theory, after you've stirred milk into your coffee the two liquids could spontaneously separate out again and yet this has never been observed to happen. Why not?

Around 1876, Austrian physicist Ludwig Boltzmann used the kinetic theory to resolve the so-called 'reversibility paradox' in physics. He realized that there must be many more disordered states for a system than there are ordered states; therefore, random interactions will inevitably lead to greater disorder. There are a colossal number of ways for the molecules of milk and coffee to be mixed together, but only one way for them to be separate.

Boltzmann solved the reversibility paradox by determining that the second law was about probabilities. All of the countless atoms and molecules that make up an object are in constant random motion. There is a vanishingly small, but not absolutely impossible, chance that the molecules in your café au lait will all move in just the right direction to separate out. But the chances of this happening are just so utterly improbable that the mixing of milk and coffee is effectively irreversible. Albert Einstein thought Boltzmann's theory was 'absolutely magnificent'.

THE ARROW OF TIME

The idea of an 'arrow of time', pointing from the past to the future, was first introduced by the astronomer Sir Arthur Eddington. A consequence of the second law and the ideas of entropy and irreversibility is that the universe, as a closed system, must eventually approach a state in which its entropy has the highest possible value. The universe proceeds inexorably from a state of low entropy (order) to a state of high entropy (disorder). It will reach a state of equilibrium in which all forms of fuel will have been expended and all available energy will have been converted into heat, the temperature will be uniform throughout the cosmos, with no prospect of heat flowing from one place to another and therefore no possibility of any work being accomplished. In another phrase made popular in the 1930s by Eddington, the universe will have entered into a final state of 'heat death'.

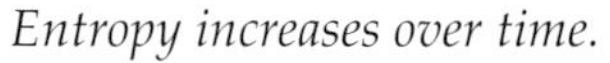

Entropy increases over time.

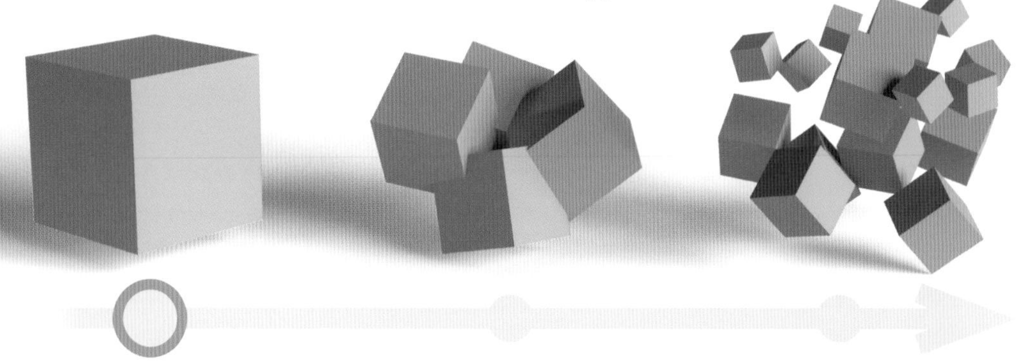

13

Molecules in motion

Kinetic theory

That the properties of matter, particularly gases and liquids, depend on the behaviour of the atoms and molecules that make it up is now accepted scientific fact, but it was for a long time hotly disputed and slow to find favour with researchers.

In 1738, Swiss mathematician and physicist Daniel Bernoulli put forward a kinetic, or movement, theory of gases. He imagined a movable piston inside a cylinder filled with tiny particles of gas moving around randomly. Pressure was the result of the impact of the particles colliding with the piston. The smaller

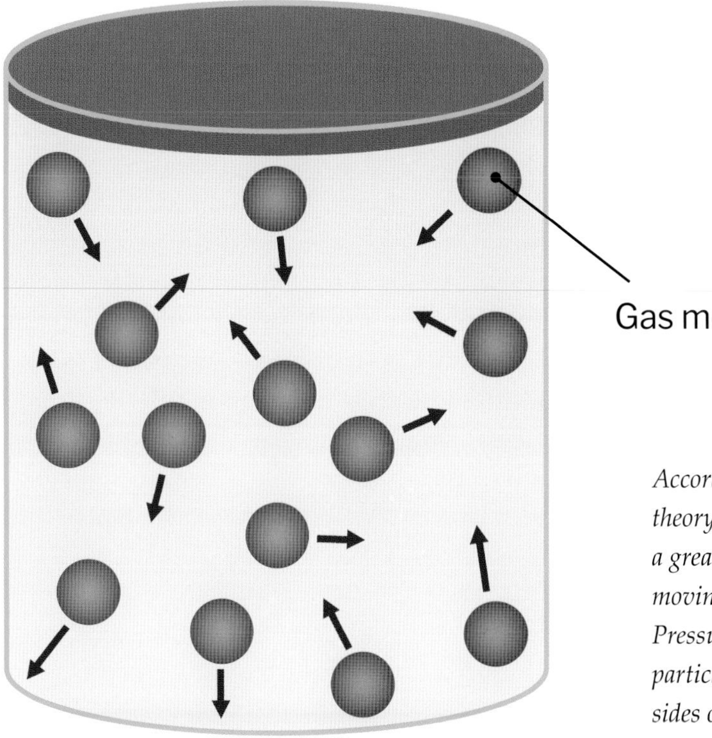

According to the kinetic theory a gas consists of a great many randomly moving particles. Pressure is a result of the particles impacting the sides of the container they are held in.

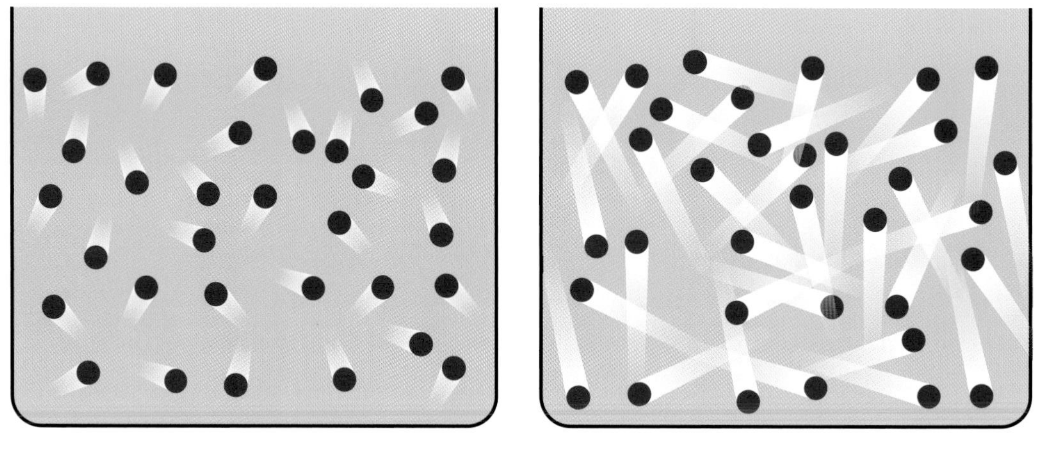

Heat is a measure of how fast particles are moving; the higher the temperature the faster the particles move.

the enclosed space, or the greater the number of particles, the more frequent will be the collisions and so the greater the pressure will be.

The gas laws, for example Boyle's law, which states that for a fixed mass of gas at constant temperature the volume is inversely proportional to the pressure, are easily explained by Bernoulli's kinetic theory. The pressure of a gas depends on the frequency with which the molecules strike the surface of the container. Compressing the gas to a smaller volume means that the same number of molecules are now acting against a smaller surface area, so the number striking per unit of area, and thus the pressure, is now greater.

Bernoulli's kinetic theory introduced the idea that heat or temperature could be identified with the kinetic energy of particles. According to kinetic theory, heat measures the motion of atoms. The more agitated the atoms, the greater the heat.

Bernoulli also showed that pressure is proportional to the kinetic energy of the particles. Increasing the temperature increased the kinetic energy of the particles. Since the frequency of impacts is proportional to the speed of the particles and the force of each impact is proportional to the particle's momentum, this explained why increasing the temperature also increased the pressure.

MAXWELL AND THE DISTRIBUTION FUNCTION

In 1859 physicist James Clerk Maxwell realized that trying to analyse what was happening with the moving gas particles using Newton's laws of motion was a hopeless task. There were just

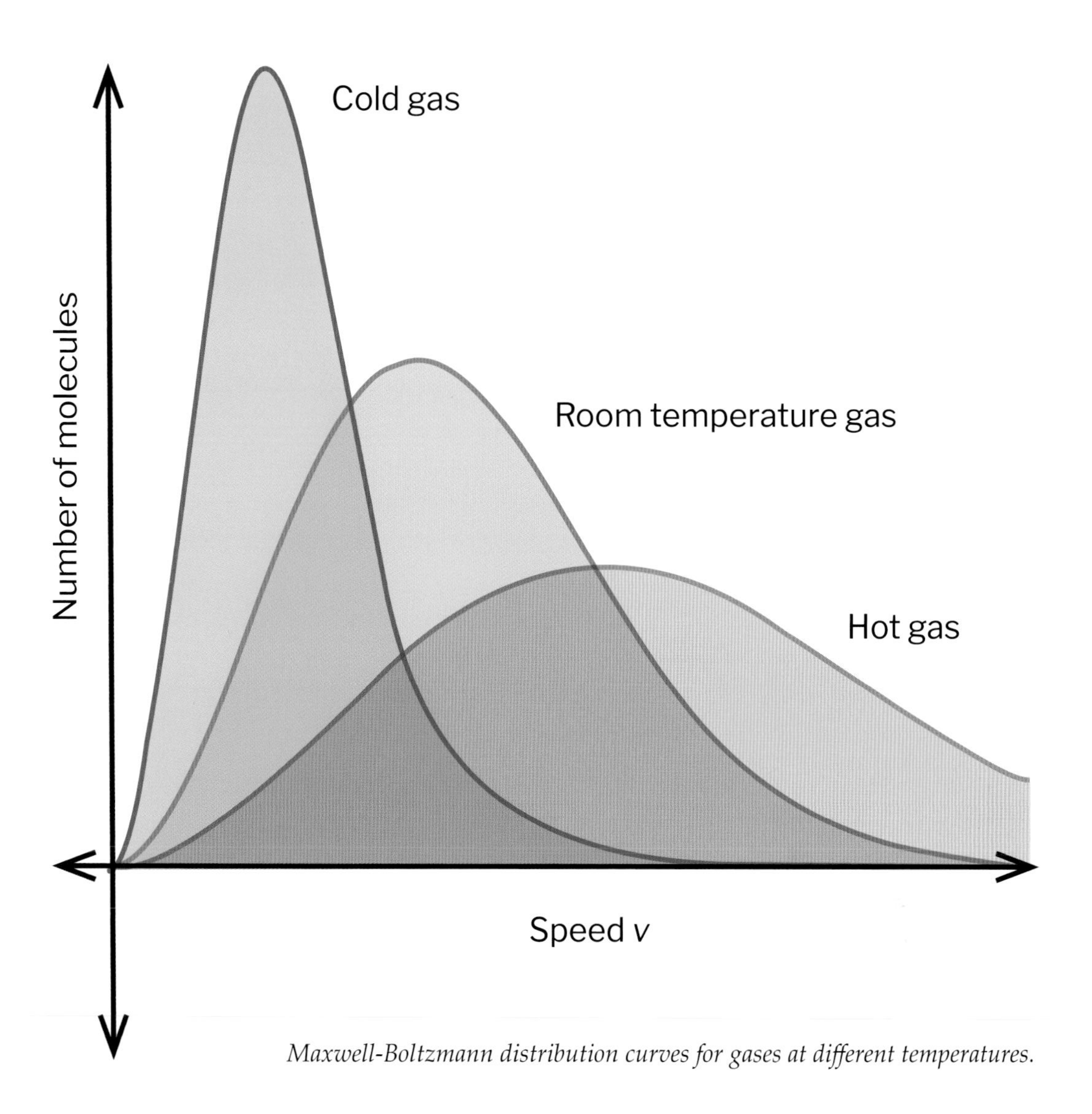

Maxwell-Boltzmann distribution curves for gases at different temperatures.

too many variables involved. Maxwell saw that there was no need to describe the movement of individual molecules, but to have an understanding of their average distribution. It wasn't necessary to have knowledge of the position and velocity of every single molecule at every instant of time, but just to be able to say what percentage of the molecules are in a certain part of the container, and what percentage have velocities within a certain range, at a particular time, a determination called the distribution function. The distribution function is independent of time for a gas in thermal equilibrium.

The kinetic molecular theory

1. Gases are composed of a large number of particles in a state of constant, random motion.
2. These particles move in straight lines until they collide with another particle or the walls of a container.
3. When they collide, they behave like hard, spherical objects. Collisions between particles or between particles and the walls of a container are perfectly elastic. None of the energy of a gas particle is lost in the collision.
4. Almost all of the volume occupied by a gas is empty space. The particles are much smaller than the distance between them.
5. No force of attraction acts between gas molecules or between the molecules and the walls of any container.
6. The average kinetic energy of the gas particles is solely dependent on the temperature of the gas.

Brownian motion

In 1827, botanist Robert Brown was studying the particles contained in grains of pollen. He noticed that these particles were in constant motion, and that this motion didn't appear to be caused by currents in the fluid they were in. The movement had nothing to do with the substance being alive or dead – dust particles jiggled just the same. So, what was causing it?

Half a century later, the kinetic theory provided an explanation. If the molecules in a fluid were in constant motion, then a particle suspended in the fluid would be bombarded from all sides by the moving molecules, causing it to move around.

Maxwell also calculated the average distance travelled by molecules between collisions (known as the mean free path) and the number of collisions that could be expected to take place at a given temperature. The higher the temperature, the faster the molecules moved, and the greater the number of collisions that took place. The distribution function is independent of time for a gas in thermal equilibrium. In 1871, German physicist Ludwig Boltzmann extended Maxwell's function to include the distribution of energy among the molecules

14

Force of attraction

Magnetism

People have been aware of the mysterious phenomenon of magnetism for millennia. From divination to direction finding, from toys to sticking notes to the fridge, the invisible force of the magnet has been endlessly fascinating. But what is it?

Writings from China and Greece dating back over 2,500 years describe the attractive power of lodestone, a form of magnetite, a naturally occurring magnetic mineral. With no visible means of doing so, it could draw iron towards it over a distance.

The magnetic attraction of lodestones was a mystery that was hard to explain. Ideas ranged from the belief that magnets

Lodestone, or magnetite, is a naturally occurring material with magnetic properties.

The Chinese first used a form of compass, not for navigation, but to determine the most propitious alignment for their homes.

had a soul, to the notion that the lodestone emitted invisible particles that carved out a void in space that other objects rushed to fill. The 1st-century CE Roman writer Pliny asked: 'What phenomenon is more astonishing? Where has nature shown greater audacity?'

That iron could itself be magnetized by stroking it with a lodestone was an important early discovery. A magnetized needle could be used to make a compass. English theologian Alexander Neckham mentions in 1190 that sailors from the East were using magnetized needles for navigation, the first reference in Europe to the use of compasses. Christopher Columbus carried one on his voyage of discovery to the Americas in 1492 and the compass was soon aiding the expansion of European empires across the world.

THE POLES OF PEREGRINUS

Though the magnet had undoubted practical purpose, the physical mechanism of magnetism was poorly understood. Thirteenth-century French engineer Pierre de Maricourt, known as Petrus Peregrinus, produced one of the first comprehensive accounts of magnetism. His experiments included methods of determining magnetic polarity and the effects of magnetic attraction and repulsion. By observing how a needle aligned on the surface of a spherical magnet he discovered that the pattern of lines converged at two points opposite each other on the sphere – the north and

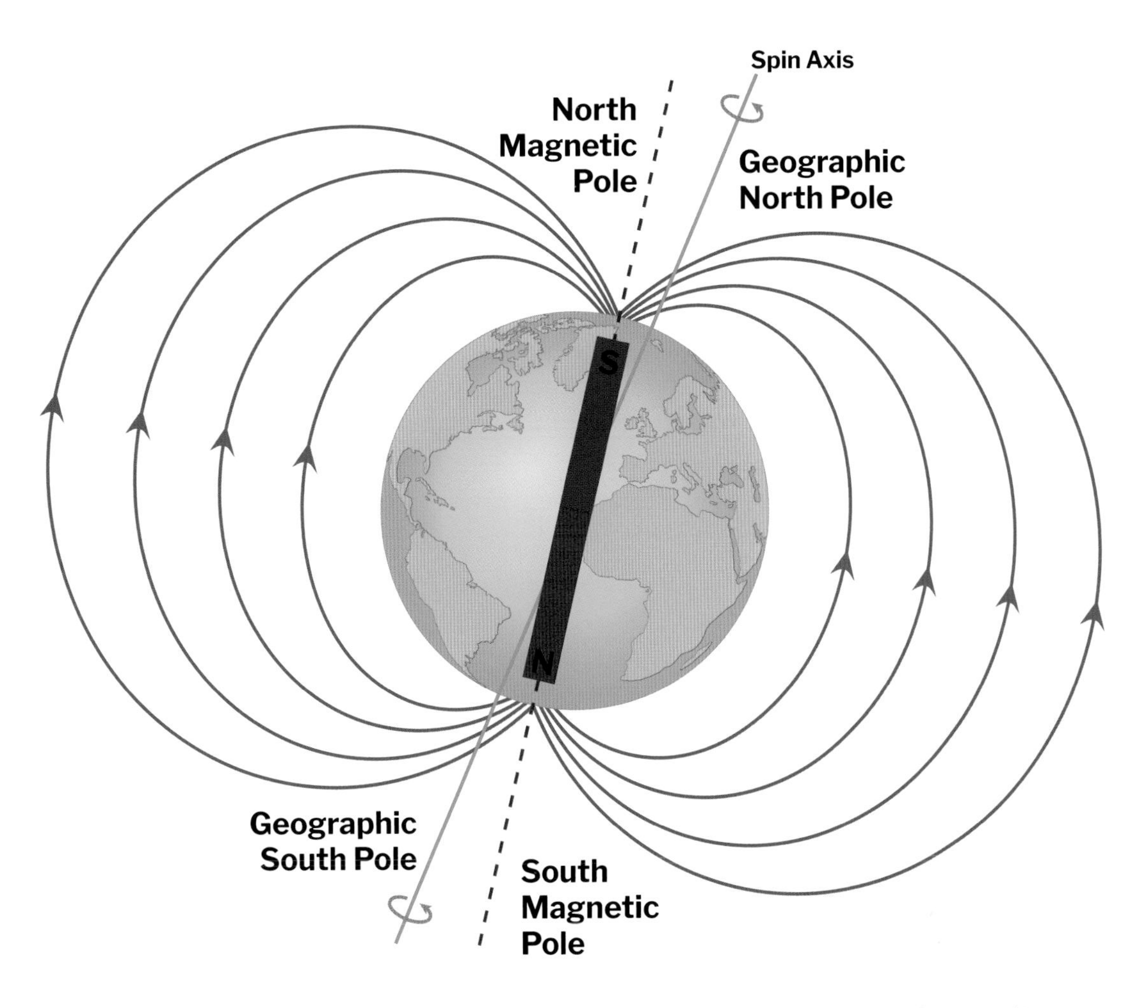

The poles of Earth's magnetic field, generated by electrical effects in its iron core, does not align with the planet's geographical poles.

south poles of the magnet. He was the first to observe that opposite poles attract whereas like poles repel and noted that when a magnet was broken in two each half had its own north and south poles.

MAGNETIC EARTH

At some point compass-steering navigators realized that their compass needles were not pointing exactly towards geographic, or true, north. This difference in the direction of magnetic north from that of geographic north, known as the magnetic declination, varies across the surface of the earth. It was also discovered that the north-pointing end of a magnetized needle that was free to move vertically also pointed downwards, a phenomenon called magnetic dip, or inclination. The force that attracted the needle northwards was also drawing it towards the earth.

Scientists struggled to come up with an explanation for the fact that the earth's magnetic north pole, and its geographic north pole were not in the same place. Evidence that the difference between the magnetic and geographic poles varied over time complicated matters. It meant that compass bearings became inaccurate over a period of decades and had to be recalibrated. Why this should be was a mystery for centuries.

Earthquake studies have shown that the earth has a layered structure and that differences in the rate of spin between the solid inner core and the liquid outer core may be responsible for generating the earth's magnetic field. As one moves against the other, electric currents are induced. Electric currents generate magnetic fields, so the theory is that circulating electric currents in the earth's molten metallic core are the cause of the magnetic field. Because these forces are constantly changing, the strength of the field varies over time, causing the location of the magnetic poles to shift.

WHAT IS MAGNETISM?

Everything is made up of atoms, and each atom has one or more electrons surrounding a central nucleus. Each electron can be thought of as a tiny

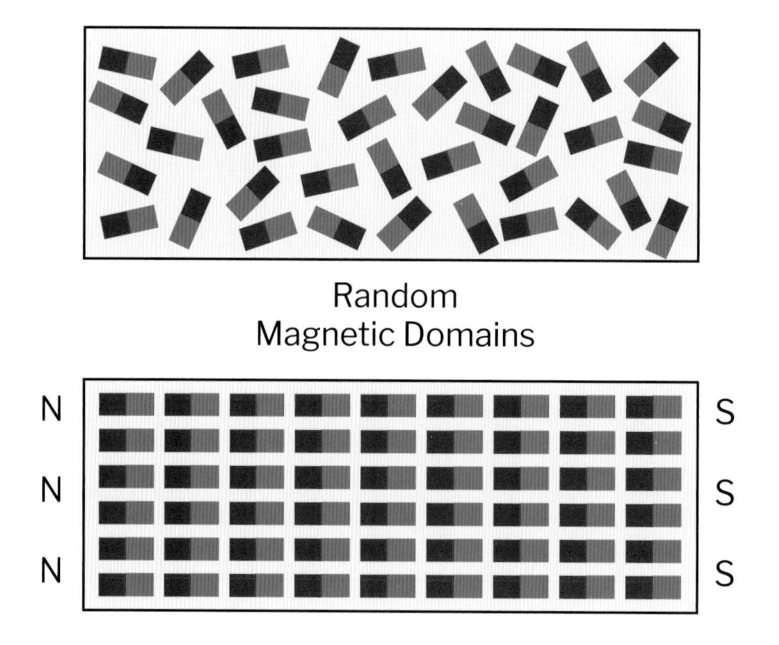

The phenomenon of magnetism may be the result of aligning the tiny magnetic fields of individual atoms.

electric charge and a moving electric charge produces a magnetic field. This means that the atoms in an object act like many tiny magnets. The magnetic field surrounding an atom has a specific orientation or direction, called the atom's **magnetic moment**. In most materials, the magnetic moments are oriented in random directions and effectively cancel each other out, which means that the object will not act like a magnet. However, if all or most of the magnetic moments are aligned in the same direction, the object creates a magnetic field around itself. In ferromagnetic materials, the atoms form structures called **domains**, regions where groups of **magnetic moments** naturally align in the same direction.

The little earths of William Gilbert

In 1600, William Gilbert published *De Magnete* (*On the Magnet*), one of the first great works of physics and experimental science. Gilbert's careful use of experiments marked him out as one of the forerunners of the modern scientific method. He carried out experiments using a spherical lodestone, which he called a terrella, or 'little earth', as a model for the earth. (Present-day scientists have employed terrellas inside vacuum chambers to, for example, mimic the effect of the earth's magnetic field on cosmic ray particles and the solar wind.)

William Gilbert.

Gilbert made the bold assertion that the whole planet earth was a giant lodestone and suggested that magnetism was what caused the earth to rotate, and that magnetic attraction was responsible for holding the moon in orbit around the earth, and for the moon's influence on the tides.

15

A flash of lightning

Static electricity

Humans have been aware of electrical phenomena throughout history. Early civilizations believed that lightning bolts were flung across the sky by the gods and lightning strikes may well have provided humans with their first source of fire. Understanding, still less control of, electricity was a long time coming.

Have you felt the crackle when you pull a jumper over your head, been startled by a sudden flash of lightning or tried to stick a balloon to the wall after rubbing it on some wool?

These are all the result of static electricity. Normally, the atoms in a material contain equal numbers of positively charged protons and negatively charged electrons balancing each other out so the material is

A balloon given a negative charge of static electricity by rubbing repels the negative charges in the hair, giving it a small positive charge. The opposite charges attract, pulling the hair towards the balloon.

A triboelectric series ranks various materials according to their ability to gain or lose electrons.

electrically neutral. Static electricity is the result of an excess (or deficit) of electrons. When something is 'charged' with static electricity, this means that the number of negatively charged electrons present is greater than the number of positively charged protons. For some materials, rubbing them together can cause a very small number of electrons to cross from one of the materials to the other. Materials that can lose (or gain) electrons in this way are called triboelectric. The material that loses electrons will have an overall positive charge and the material gaining the electrons a negative charge of the same magnitude.

Charged objects exert forces on each other. Similar charges repel and opposite charges attract. If a negatively charged object is brought close enough to a positively charged one, electrons will stream from one to the other, neutralizing the static charge. For example, we can acquire a static charge walking over carpets or rubbing against furnishings that may flash painfully away when we touch a door handle.

STANDING CHARGES

William Gilbert, who also investigated magnetism, studied electrical phenomena using an instrument called a 'versorium' (from the Latin meaning 'to turn about'), a light metal needle, balanced on a pinhead at the midpoint that was a sensitive detector of electric properties. Gilbert compiled a list of over 20 materials, including glass, sapphire, sealing wax and amber, that could be electrified by rubbing – an effect he called 'electricus'. He observed that, unlike a magnetized object, an electrified object had no poles, and could be blocked by a sheet of paper.

Seventeenth-century German scientist Otto von Guericke invented the first electrical machine. This consisted of a sphere of sulphur, about the size

Amber attractions

Amber is the fossilized resin of a now extinct coniferous tree, mostly found in the Baltic region of northern Europe. Its warm yellow colour and attractive appearance make it greatly prized for jewellery. The Greeks called amber 'elektron', the origin of our word electricity. The 6th-century BCE Greek philosopher Thales described amber's ability to attract light objects, such as hair and feathers, when it was rubbed with wool and later writers described how spinners in Syria would place amber on the ends of their spindles, calling it 'the clutcher'. As the wheel spun the amber became charged and attracted stray bits of wool and chaff.

of a child's head, with a wooden rod through the middle. The ends of the rod were placed on supports allowing the sphere to be rotated easily. Rotating and rubbing the sphere electrified it, so that it attracted chaff, feathers and similar small objects. Von Guericke also observed the phenomenon of electrical conduction, noting that a thread attached to the sphere would show electrical attraction at its far end. The rotating sulphur sphere soon became the standard way of producing electricity and remained so for over a century.

By the middle of the 18th century electricity generators had improved to the point that they were becoming dangerous. In 1745, Dutch physicist Pieter van Musschenbroek invented the Leyden jar, a device for storing electricity. This was a glass jar filled with water with metal foil lining the inside and around the outside. Today, we might think of it as an early form of capacitor. A typical design has a metal rod emerging from the cap of the jar, connected to the inner lining by a

Otto von Guericke invented a machine that generated an electric charge by rotating a sulphur sphere.

chain. The rod is charged with electricity by touching it to a charged-up glass or sulphur sphere, for example. Several jars could be connected in parallel to increase the amount of charge stored. Curious to see if electricity was indeed being stored, van Musschenbroek touched the inside and outside of the jar simultaneously, and succeeded in electrocuting himself, luckily not fatally.

Keen observers noted the similarity between the spark of a static discharge and the bigger flash of a bolt of lightning. In 1752, American diplomat and scientist Benjamin Franklin decided to put the similarity to the test. Famously, Franklin used the string of a kite to charge up a Leyden jar during a thunderstorm, thus proving that the thundercloud was electrically charged. Lightning comes about through the same physical processes that make your hair stand on end when you pull a jumper off. It is just happening on a much

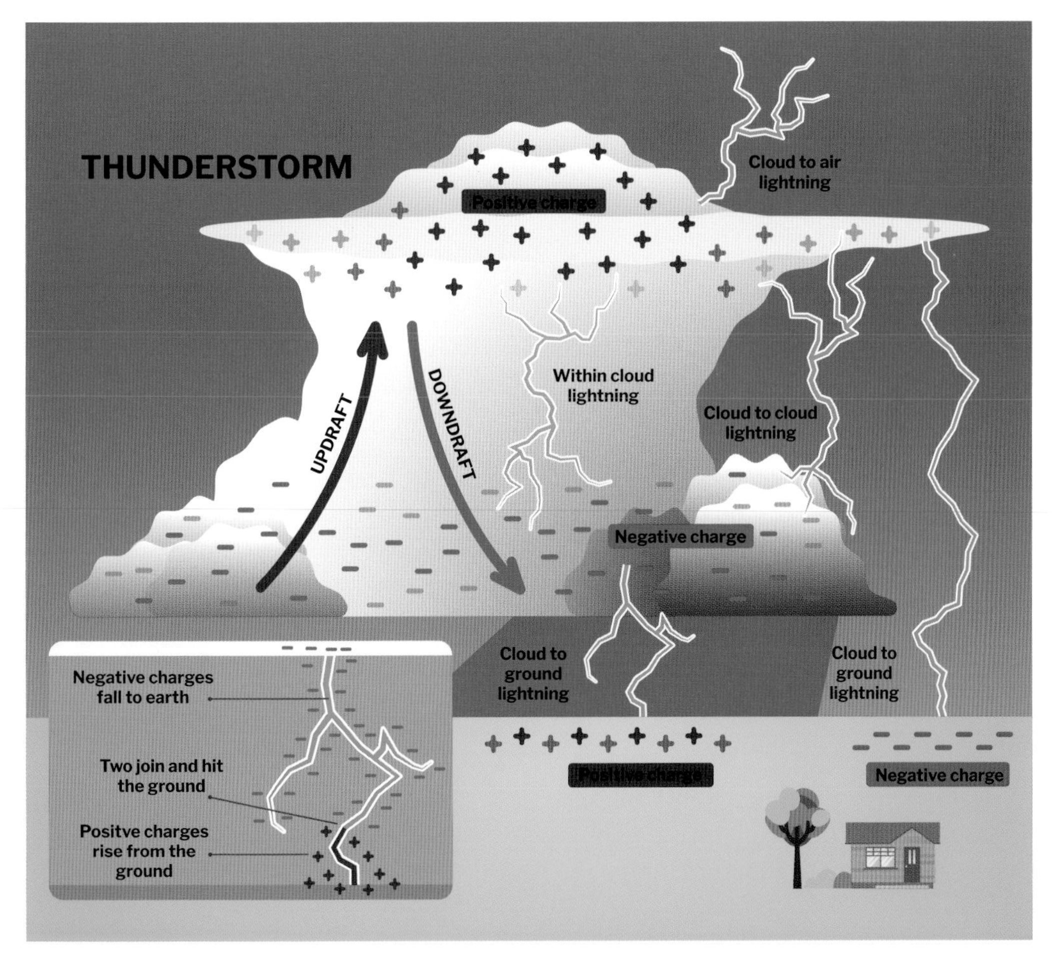

Lightning is the result of a huge build-up of static charge inside a thundercloud.

larger scale. As a thunderstorm builds, circulating air currents cause water droplets in the atmosphere to acquire electric charge. The water droplets at the bottom of a thundercloud steadily become more negatively charged while at the same time the ground beneath builds an equal positive charge. Because air is a very poor conductor, the charge differential continues to increase until it becomes powerful enough to cause the air molecules to ionize. A channel of ionized air, called a leader, reaches down from the base of the cloud to the ground. When the charged column of electrons makes contact with the ground the accumulated electric charge flashes down as a lightning bolt. The massive burst of electrical energy can heat the surrounding air to 30,000°C (54,000°F), resulting in a flash of brilliant light and causing an explosive expansion of the air heard as a crash of thunder.

COULOMB'S LAW

French physicist Charles Coulomb carried out the first quantitative study of electrostatic force in the 18th century. He invented a sensitive torsion balance that allowed him to measure accurately the force produced by an electric charge. Coulomb was able to demonstrate that electric charges obey a law:

$$F = k\, q_1\, q_2 / r^2$$

where: k = the electrostatic constant, q1 and q2 are the two charges, and r = distance between the charges.

Compare this with Newton's law, which simply substitutes the gravitational constant G for the electrostatic constant k and mass for charge. Both forces decrease proportionally as the inverse square of the distance, and both depend directly on the charge (or mass) of the two objects exerting force on each other. The electrostatic force is enormously more powerful than the gravitational force.

Conservation of charge

The law of the conservation of charge is one of the fundamentals of physics. It states the net charge of an isolated system will always remain constant. If two objects in an isolated system have a net charge of zero, and one object donates negatively charged electrons to the other, the object receiving the electrons will be negatively charged and the object donating the electrons will have a positive charge of the same magnitude. The total charge of the system remains zero. Overall, this applies to the entire universe; every charged particle has a corresponding particle of equal and opposite charge.

16

Current events

Electrical energy, current and resistance

Static electric charges can be very powerful, but they generally stay in one place and can't do anything useful. All of the many electric devices we use today depend on a flow of electric current to make them work.

Inventions such as the Leyden jar gave scientists the means to store electricity, but the charge could only be released all at once in a single spark. There was no way to release the energy gradually in a steady flow and do something useful with it.

Around 1780, Italian scientist Luigi Galvani had begun investigations into the effects of electricity on dissected frogs, discovering that he could make the dead frogs' legs kick by applying a spark of electricity or by touching them with different metals. Galvani concluded that a frog's body contained a type of electrical fluid, naming it animal electricity.

VOLTA'S PILE

Self-taught Italian physicist Alessandro Volta was intrigued by Galvani's discovery. He suspected that the answer lay not with the frogs but with the metals Galvani used. He thought that the contact between the different metals was what was generating the electricity. There being no suitable instrument available at the time to detect the electrical output, so Volta relied on popping various combinations of metal into his mouth to see what happened.

Finding zinc and copper to give the best results, Volta built a vertical pile of alternating zinc and copper discs, separated by circles of cloth soaked in brine. When he connected a wire to each end of the pile he found that he had created a flow of electricity. When he touched the wires, he felt a slight shock, which gradually increased in force the more discs he added to the pile. For the first time, it became possible to produce a steady supply of electrical power. After Volta reported his discovery in 1800 the electric pile, the forerunner of the battery, became a huge success, spurring investigations into the nature of electricity.

Volta's pile was a simple electrochemical cell, a device that

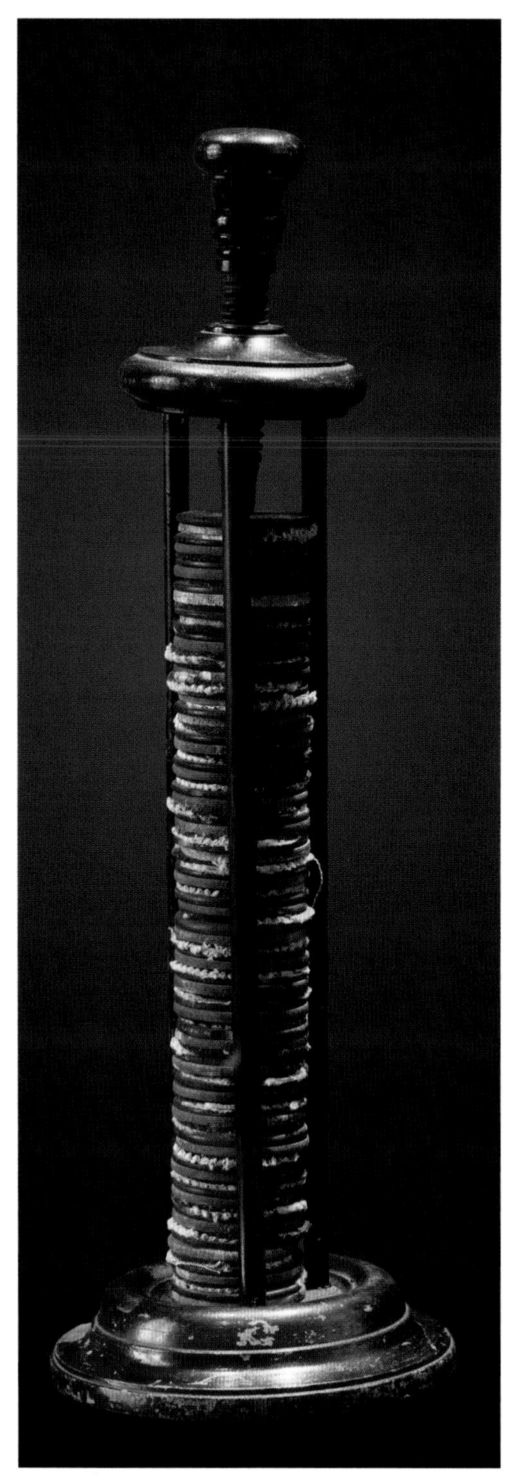

A voltaic pile.

converts chemical energy into electrical energy. It consists of two metal electrodes, the anode (zinc in Volta's pile) and the cathode (copper), separated by an electrolyte (the brine). The metal at the anode gives up electrons (an oxidation reaction) and these flow through the electrolyte to be taken up by the metal at the cathode (a reduction reaction). School textbooks reinforce this sequence through the 'oilrig' mnemonic – 'oxidation is loss, reduction is gain'.

Volta's pile only supplied current for a short time before the chemical reactions driving it came to a halt, but the principle had been demonstrated and later developments would see batteries with greatly improved longevity.

CURRENT FLOW

Volta's battery demonstrated that an electric current can only flow if it has something, a conductive material, to flow through. Current electricity depends on the flow of electrons. Some substances, such as metals, are generally very good conductors of electricity. Some of the electrons in a metal are free to move around. In normal circumstances these movements are random, but when there is a potential difference or imbalance in charge applied between two points the electrons all begin to move in the same direction. Current flows from higher potential to lower, by convention from positive to negative. Although the electrons themselves move relatively slowly, in a copper wire for example

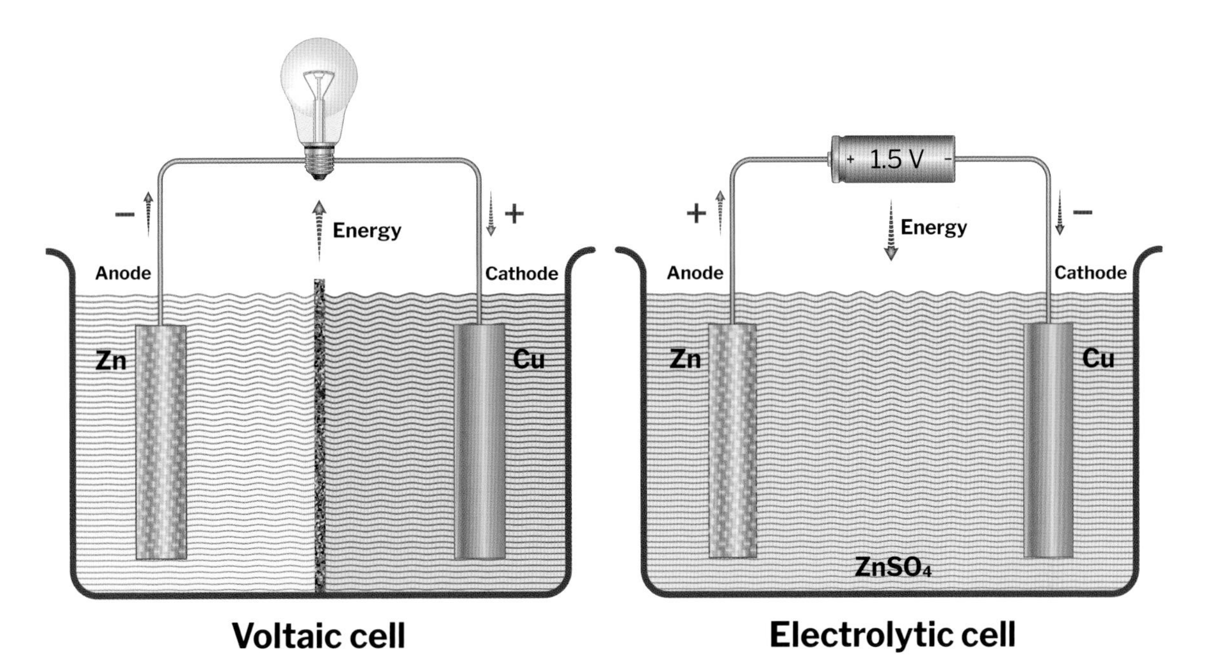

Electrochemical cells are either Voltaic (galvanic) or electrolytic. In galvanic cells a spontaneous flow of electrons from the anode to the cathode produces an electric current. Electrolytic cells require an input of electrical energy to cause the electrons to flow.

this electron drift is only about a fraction of a millimetre per second, the transfer of energy through the circuit is close to the speed of light. This is because the interactions between the moving electrons via the electric field due to their charge and the magnetic field created by their movement sets up an electromagnetic wave.

Materials such as ceramics which have few or no free electrons are poor conductors. Such materials are termed insulators. Anything in an electrical circuit that uses up electrical energy reduces the flow of the current. This is called resistance. Just as the flow of water is reduced when it passes through a narrow pipe so a long, narrow wire increases resistance in a circuit. German physicist Georg Ohm formulated a law in the 19th century relating voltage to current: Current = Voltage / Resistance. Dividing the voltage (in volts) by the current (in amps) gives the resistance of the conductor (measured in ohms, after Ohm). Ohm's law is not universal and does not hold for all conductors.

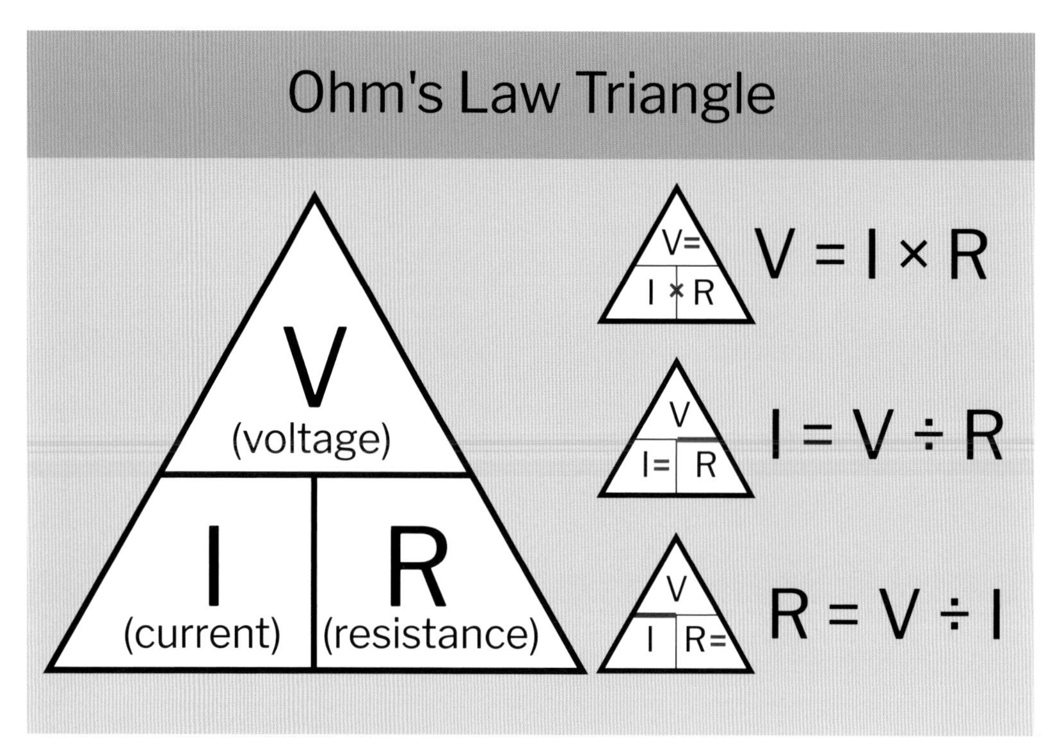

Ohm's law sets out the relationship between voltage, current and resistance.

Current and potential

Current is simply the rate at which the charge flows. A large current means a lot of charge flowing every second, transferring a lot of energy. The flow of electric current from a battery or the sudden discharge from a lightning bolt or a Leyden jar are the result of electric potential. The electric potential at one point is always measured relative to the potential at another point – it is not an absolute quantity. An imbalance in the electric potential gives rise to a potential difference, measured in volts in honour of Volta, and what in general terms is referred to as voltage. The potential difference can be thought of as the 'push' that gets the current flowing.

The unit of electric charge is the coulomb (C). The rate of flow of electric charge is measured in amps. A current of 1 amp means that 1 coulomb of charge, or about 6 trillion electrons, passes through the circuit every second. When 1 coulomb of charge moves through a potential difference of 1 volt it transfers 1 joule of energy.

17

Two become one

Electromagnetism, motors and generators

Up until 1820, most scientists agreed that electricity and magnetism, although similar in some ways, were fundamentally different forces. The discovery that they were, in fact, intimately related and one force could generate the other was a groundbreaking insight that would change the world forever.

By the end of the 18th century, knowledge of magnetic and electrical phenomena was growing, although most scientists still believed them to be separate forces. It was a chance discovery by Danish physicist Hans Christian Oersted in 1820 that led to the realization that the two are intimately linked. While giving a lecture to students, Oersted observed the deflection of a compass needle when an electric current was switched on and off, the first time that one force had been seen to affect the other.

Hans Christian Oersted.

A CIRCLING FIELD

French physicist André-Marie Ampère showed that an electric current flowing through a wire produces a magnetic field circling around the wire. If you make a 'thumbs up' gesture with your right hand, your thumb points in the direction of the current and your curled fingers indicate the direction of the magnetic

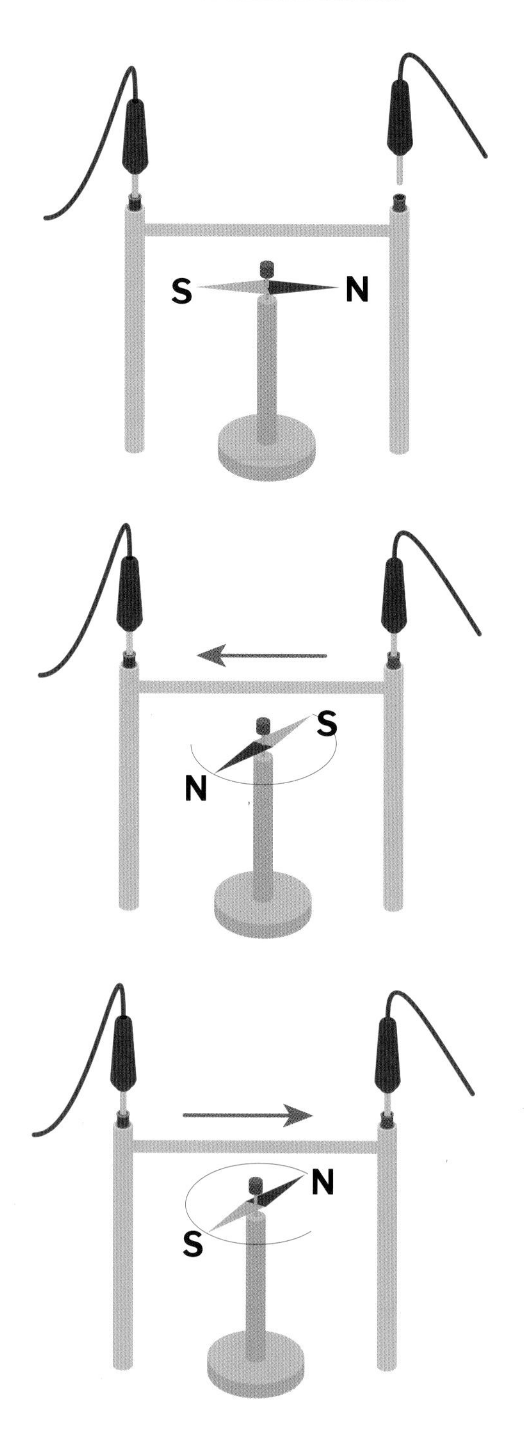

A compass needle is deflected by an electric current.

field. The strength of the magnetic field is proportional to the strength of the electric current that generates it. Reversing the flow of the current reverses the direction of the magnetic field. Parallel wires carrying an electric current will either attract or repel each other depending on the direction the current is flowing through each one.

The magnetic field around a single wire is relatively weak but winding the wire into a coil, called a solenoid, causes the fields around each loop to reinforce each other, creating a stronger magnetic field within the coil. The greater the number of loops in the coil, the stronger the field produced. An iron core inside the solenoid further strengthens the field of the electromagnet. Whereas the strength of a permanent magnet is fixed, the electromagnet has the advantage that its strength can be varied by altering the strength of the current flowing through the solenoid. Just like a bar magnet, the coil has a north pole at one end and a south pole at the other.

FROM MAGNETS TO MOTORS

If a loop of wire is placed between opposite poles of a pair of magnets, a current passing through the wire will produce an upward force on one side and a downward force on the other, causing the loop to make a half turn. At this point the direction of the force is reversed and the loop comes to a stop. Things can be kept moving with a simple device called a commutator. This is a metal ring split

Fleming's left-hand rule

A wire carrying a current placed in a magnetic field is subject to a force resulting from the interaction of the magnetic field with the field created by the current. This force causes the wire to move, a phenomenon known as the motor effect. The force is greatest when the flow of the current is at right angles to the magnetic field. The direction of the force can be determined using Fleming's left-hand rule (named after 19th-century physicist John Fleming). Hold the thumb, index finger and second finger of your left hand at right angles to each other. The index finger indicates the direction of the magnetic field, the second finger the direction of the current, and the thumb the direction of the resultant force.

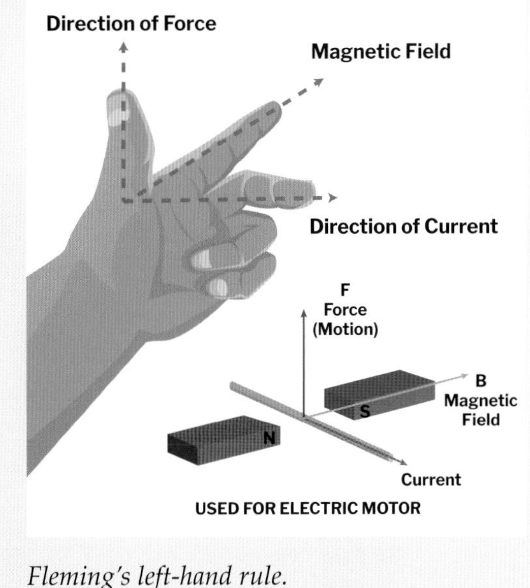

Fleming's left-hand rule.

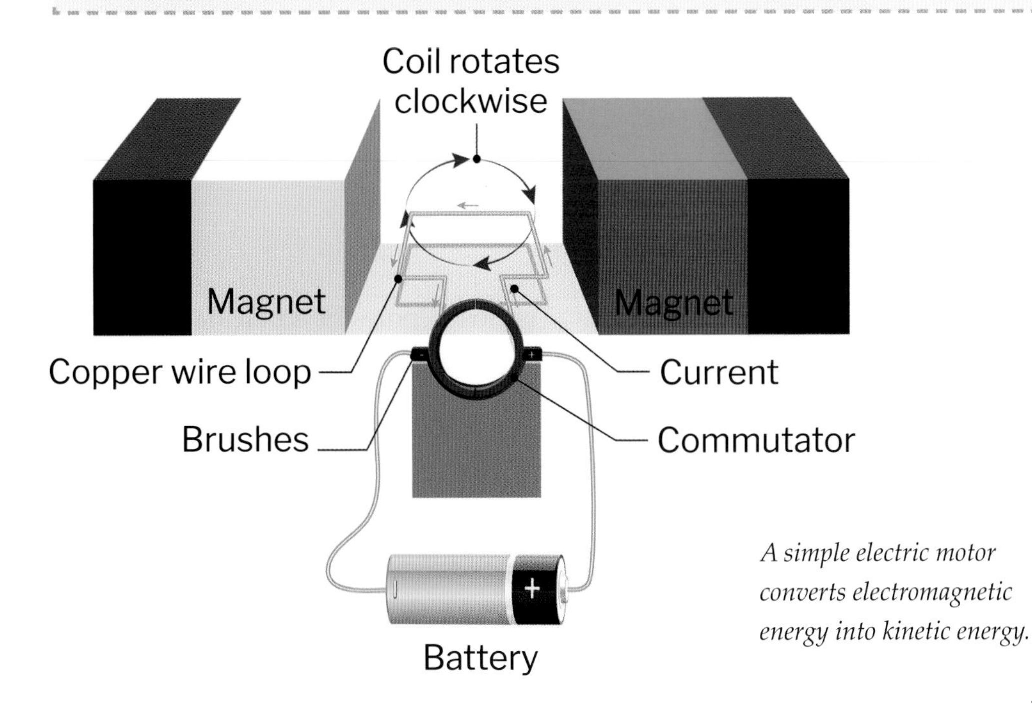

A simple electric motor converts electromagnetic energy into kinetic energy.

Fleming's right-hand rule

If the direction of movement changes, then the direction of the induced voltage also changes. The movement of the wire, the magnetic field and the direction of the current are all at right angles to each other in accordance with Fleming's right-hand rule. With the thumb, forefinger and second finger of the right hand held at right angles to each other, the thumb indicates the movement of the wire, the forefinger the direction north to south of the magnetic field, and the second finger the direction of the current.

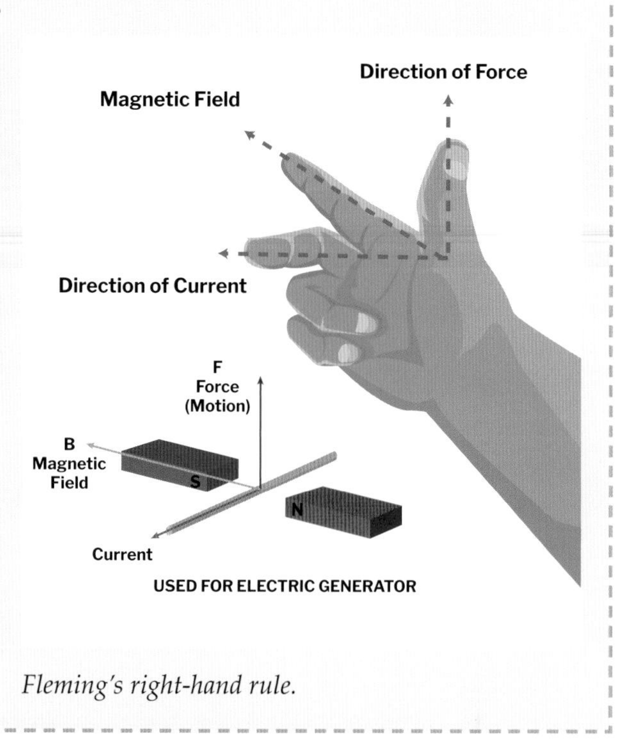

Fleming's right-hand rule.

in two which is attached to the ends of the loop. The commutator reverses the current in the loop every time it makes a half turn, resulting in the loop continuing to spin in the same direction.

From here it is a fairly straightforward step to harness the motive power of the spinning loop and produce an electric motor capable of carrying out any number of tasks. The turning force can be increased by increasing the number of loops, using more powerful magnets, or increasing the strength of the current.

ELECTROMAGNETIC INDUCTION

Just as a moving electric current produces a magnetic field, so a moving magnetic field produces an electric current. When a magnetic field is moved across a wire, or when a wire that forms part of a circuit is moved across a magnetic field, a voltage is induced in the wire, causing a current to flow. The size of the induced voltage depends on the strength of the magnetic field and the speed with which the wire is moving. This is electromagnetic induction, a discovery that changed the world forever (see chapter 18). It is the

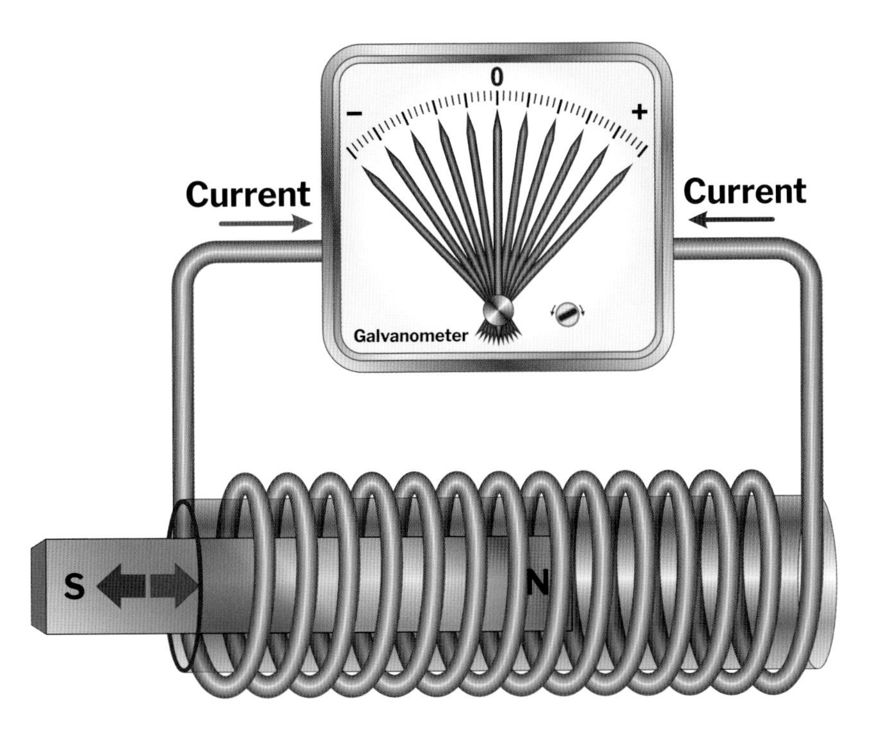

The direction of the induced current depends on the direction of movement of the magnet in the coil.

phenomenon that lies at the heart of the generators and transformers that make possible the electrical power production that drives modern technology.

GENERATORS AND TRANSFORMERS

Generators are devices that convert kinetic energy into electrical energy by electromagnetic induction. A generator that produces direct current is called a dynamo, while a generator producing alternating current is an alternator.

A dynamo has three main components – the stator, which produces the magnetic field; the armature, which is the coil rotating in the magnetic field to induce the voltage; and a split ring commutator which reverses the coil connections each half turn. This results in a direct current, one which always flows in the same direction. Increasing the strength of the stator's magnetic field, the rate of rotation of the armature, or the number of turns on the coil will all result in an increased current.

An alternator has a large coil of wire spinning inside a magnetic field. Slip rings rotate with the coil and connect it to an electrical circuit while carbon brushes ensure a continuous contact. With each half turn the coil's magnetic field reverses, resulting in the direction of

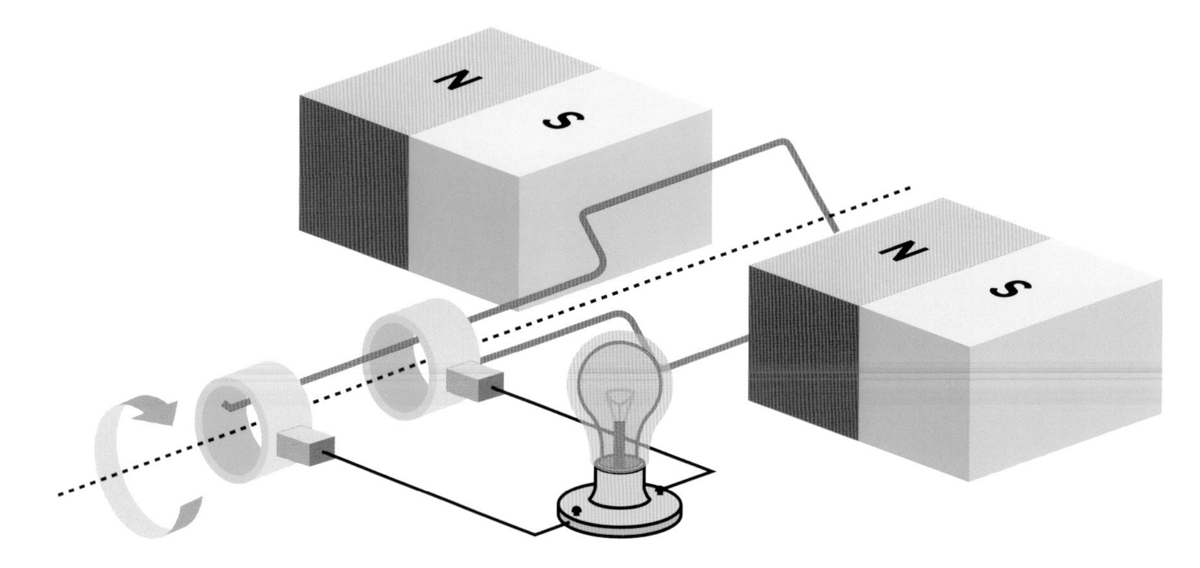

A simple dynamo converts the kinetic energy of a spinning coil into electrical energy.

the current also reversing, thus generating an alternating current, one that changes direction.

A transformer increases or decreases the voltage of an alternating current. It consists of two coils wound around a metal core. An alternating voltage is applied to one coil (called the primary coil) which sets up an alternating magnetic field in the metal core. The magnetic field in turn induces an alternating voltage in the other, secondary, coil. The size of the induced voltage depends on the number of turns in the primary and secondary coils and on the size of the voltage applied to the primary coil. Fewer loops on the secondary coil than on the primary result in a reduction, or step down, in the voltage; more loops on the secondary than the primary result in an increase, or step up, in the voltage.

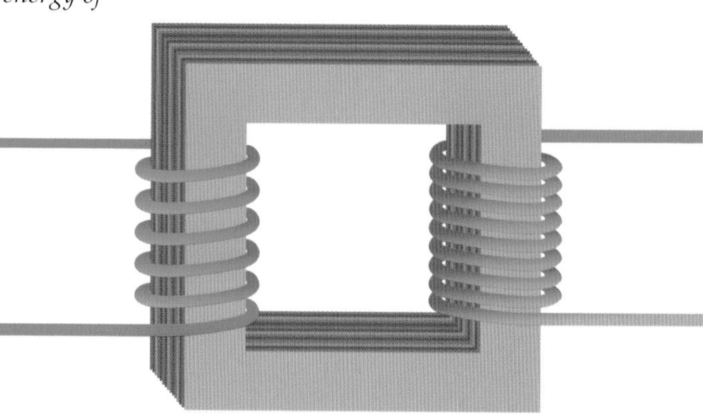

A simple transformer.

18

Lines of force

Force fields

Michael Faraday's experiments in the early 1830s into the phenomenon of electromagnetic induction, the generation of an electric field by a varying magnetic field, led him to speculate on how a force could be instantaneously transmitted between two bodies through empty space. His insights would lead to a unified understanding of electricity and magnetism.

Faraday wanted to explain how a magnet could induce an electric current in a wire without coming into physical contact with it, or an electric current make a compass needle move. To do so, he came up with the idea of an electromagnetic field. He envisaged this as lines of force, which he called 'flux lines', stretching invisibly through all of space. The patterns formed by iron filings scattered on a sheet above a magnet made the force lines visible. According to Faraday's field theory, the magnet concentrated the magnetic lines of force around itself, rather than the magnet creating the field. The magnetic force wasn't something in the magnet but in a magnetic field in the space surrounding it. The field itself exists even when there is no magnet present. The lines of force point in the direction in which the force acts and the concentration of lines corresponds to the strength of the force.

FOLLOW THE FORCE

The line of force describes the path followed by any particle within a field that is influenced by that field, for example a charged particle in an electric field or a mass in a gravitational field. The electric lines of force that represent the field of a positive electric charge in space take the form of straight lines radiating out in all directions from the charge. Since all magnets are bipolar, no magnetic monopole having ever been discovered, the magnetic field loops around from one pole to the other, emerging from the north-seeking pole and flowing into the south-seeking pole. The poles of the magnet are where the magnetic force is strongest since this is where the lines of force are most closely crowded together.

Any quantity that varies with position can be considered as a field. A magnetic field describes the region around a magnet in which magnetizable

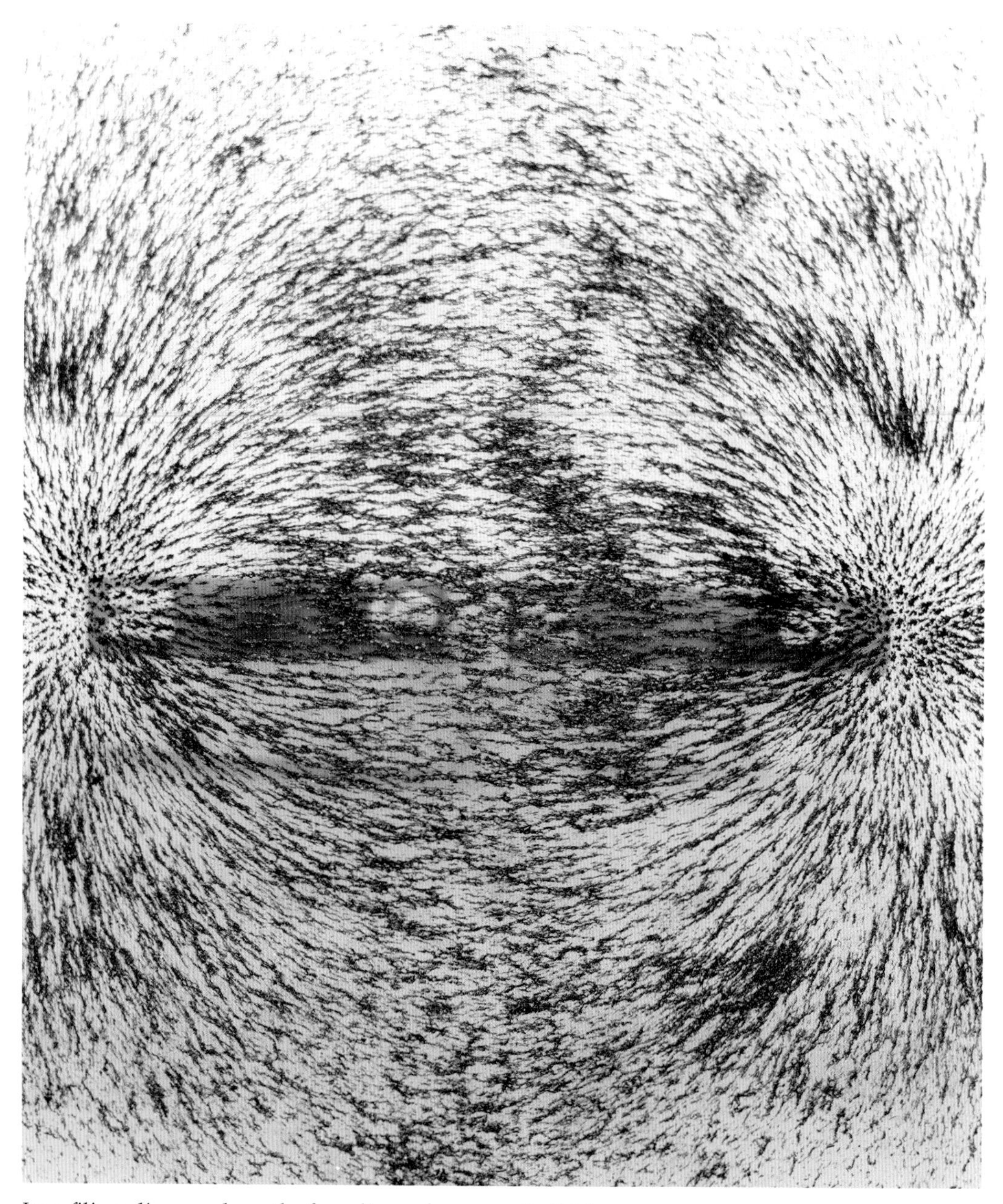

Iron filings line up along the force lines of a magnetic field.

objects, such as iron filings, feel the effects of the force. The magnitude of the force felt depends on the density of the force lines at that point. The further away from the magnet the less concentrated are the force lines and hence the smaller the force exerted. The same applies to a gravitational field extending out from a mass, which like the magnetic field has strength and direction.

GAUSS'S LAWS

By convention, in an electric field the field lines indicate the direction of the force felt by a positive charge and again the strength of the field is indicated by the concentration of the field lines. German mathematician Carl Gauss devised a law for electric fields in 1835 that related the electric field to the charges present using a mathematical calculation called the divergence. The divergence is zero if there is no charge at that point; the field lines converge where there is negative charge and flow away where there is positive charge.

Gauss's law for magnetic fields states that the divergence of a magnetic field is zero everywhere. There are no points in a magnetic field from which the field lines flow out or in. Magnetic field lines always exist as closed loops, flowing from the magnetic north pole to the south pole and through the magnet to complete the loop. Every magnet has a north pole and south pole; isolated magnetic monopoles do not exist.

Magnetic, electric and gravitational fields are all vector fields, that is to say that each point in the field has an associated strength and direction.

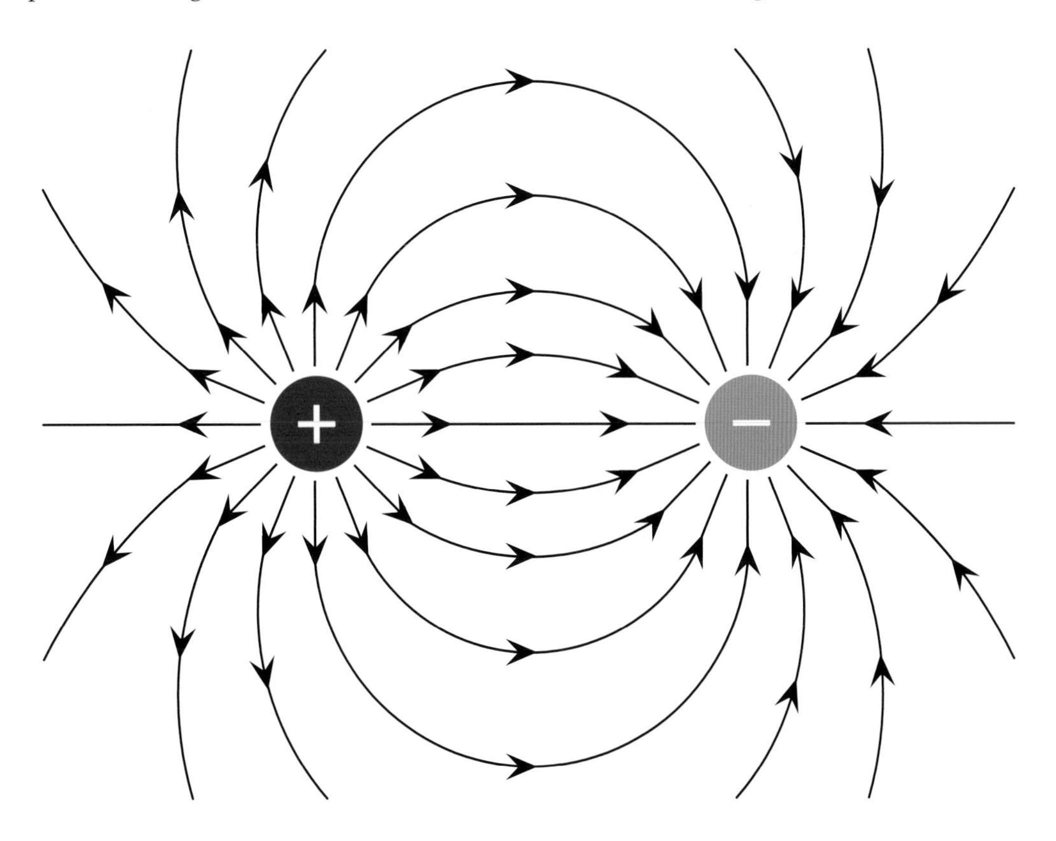

Field lines flow away from a positive charge and converge on a negative charge.

19

The waves we cannot see

Electromagnetic waves and Maxwell's equations

Nineteenth-century Scottish physicist James Clerk Maxwell is generally acknowledged as being one of the finest scientists who has ever lived. Some 20 years after Michael Faraday proposed his field theory, Maxwell took up the idea and set out to put Faraday's ideas on a firm mathematical footing.

James Clerk Maxwell was one of the great geniuses of physics.

Albert Einstein, who would overturn much of traditional physical thinking, could find no flaw in Maxwell's thoughts, describing his work on electromagnetism as 'the most profound and most fruitful that physics has experienced since Newton'.

The theory of electromagnetism was one of the greatest breakthroughs in physics and its principal architect was James Clerk Maxwell. In just four short equations, Maxwell succeeded in describing all of the electric and magnetic phenomena that had been observed and recorded by Faraday and other researchers. Maxwell's equations brought together different aspects and behaviour of both forces, showed how the two were intimately intertwined, and provided accurate predictions for future experiments. As Einstein later declared, 'it took physicists some decades to

Maxwell's equations.

grasp the full significance of Maxwell's discovery, so bold was the leap that his genius forced upon the conceptions of his fellow workers'.

Maxwell's four equations describe the electric and magnetic fields associated with electric charges and currents, and how those fields change over time. They brought together decades of experimental observations of electric and magnetic phenomena by people such as Michael Faraday and André-Marie Ampère. Maxwell established for the first time that varying electric and magnetic fields could propagate indefinitely through space in the form of electromagnetic waves as one generated the other.

WAVES IN MOTION

An electromagnetic wave can be imagined as being like two waves travelling in the same direction but at right angles to each other. One of these waves is an oscillating magnetic field, the other is an oscillating electric field. Oersted and Faraday had discovered that a varying magnetic field generates an electric field, and an electric field can generate a magnetic field. Because of this the electromagnetic field is effectively self-generating, the two fields keeping in step with each other as the wave travels along. Maxwell's proposals showed that electricity and magnetism were always bound together and that it was impossible to have one without the other.

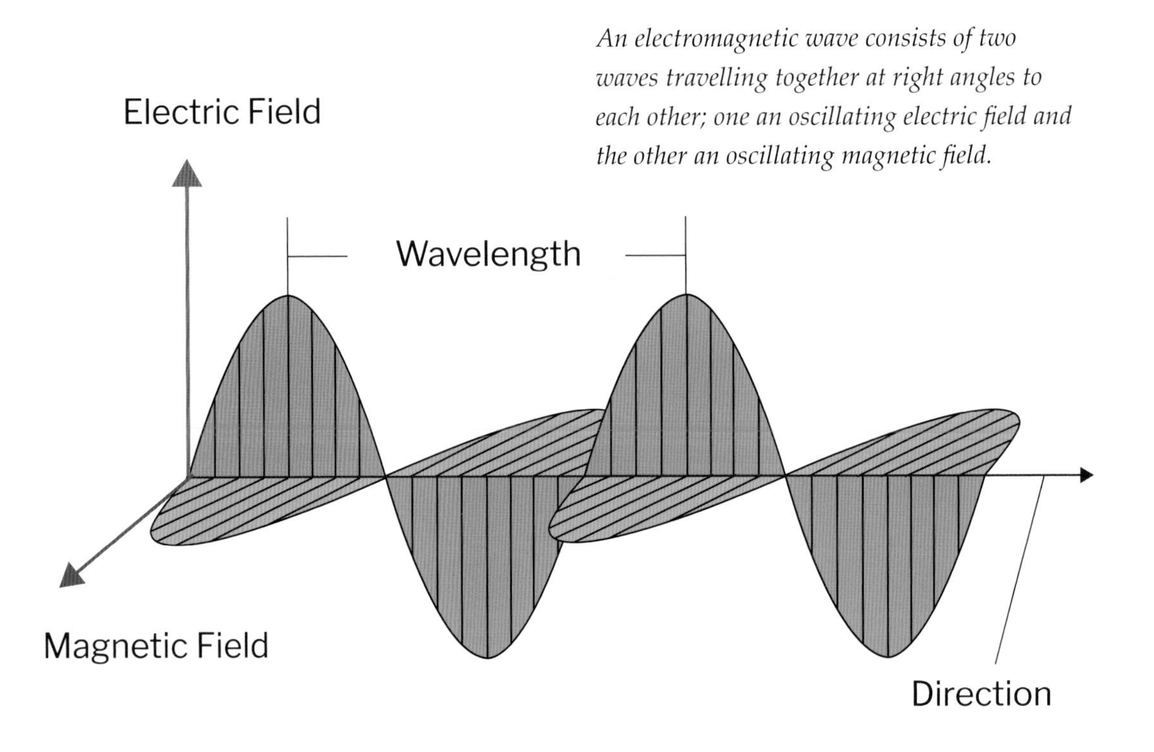

An electromagnetic wave consists of two waves travelling together at right angles to each other; one an oscillating electric field and the other an oscillating magnetic field.

Using his equations, Maxwell calculated the velocity of an electromagnetic wave as 299,792,458 metres per second (186,282 miles per second), a value that agreed with the experimental value for the velocity of light. Maxwell thought this couldn't possibly be a coincidence and had no hesitation in declaring that light itself was an electromagnetic wave. In Maxwell's equations, the speed of electromagnetic waves is a constant defined by the properties of the vacuum of space through which the waves move. It is not measured relative to anything else as any other speed would be. The nature of the universe and the behaviour of electric and magnetic fields dictate the speed with which electromagnetic waves are propagated.

THE ELECTROMAGNETIC SPECTRUM

Maxwell predicted that there should be a whole range, or spectrum, of electromagnetic waves, a prediction that proved to be accurate. Infrared and ultraviolet light, invisible to human eyes, had already been discovered at either end of the visible spectrum and scientists had demonstrated that they had the same wavelike properties as visible light. After Maxwell's death, the discovery of long wavelength radio waves and very short wavelength X-rays and gamma rays extended the spectrum further.

If light is a wave, as Maxwell

suggested, then how did it travel through the vacuum of space? Waves need some sort of medium to carry them; sound reaches your ear through the medium of the air, for example. So how does the light from the sun, say, get from there to here? Maxwell and his contemporaries concluded that light must also travel through a medium. They called it 'ether'. This mysterious substance was to all intents and purposes undetectable. Planets, moons and other solid objects travelling through space were seemingly unaffected by it. Light failed to illuminate the ether and passed through it undiminished. Yet it must fill all of space since the light from the stars reaches us from every direction. How could this ghostly substance that filled the universe be detected?

Subtle experiments aimed at detecting the ether by measuring the speed of light travelling through it in different directions found no evidence that it existed at all, a finding that shook physics.

An electromagnetic wave, produced by changing electric and magnetic fields, needs no medium to travel through. Unlike a water wave displacing water molecules as it passes through them, nothing is displaced in space by an electromagnetic wave. An electromagnetic wave is an energy-carrying disturbance travelling invisibly through space until it interacts with matter.

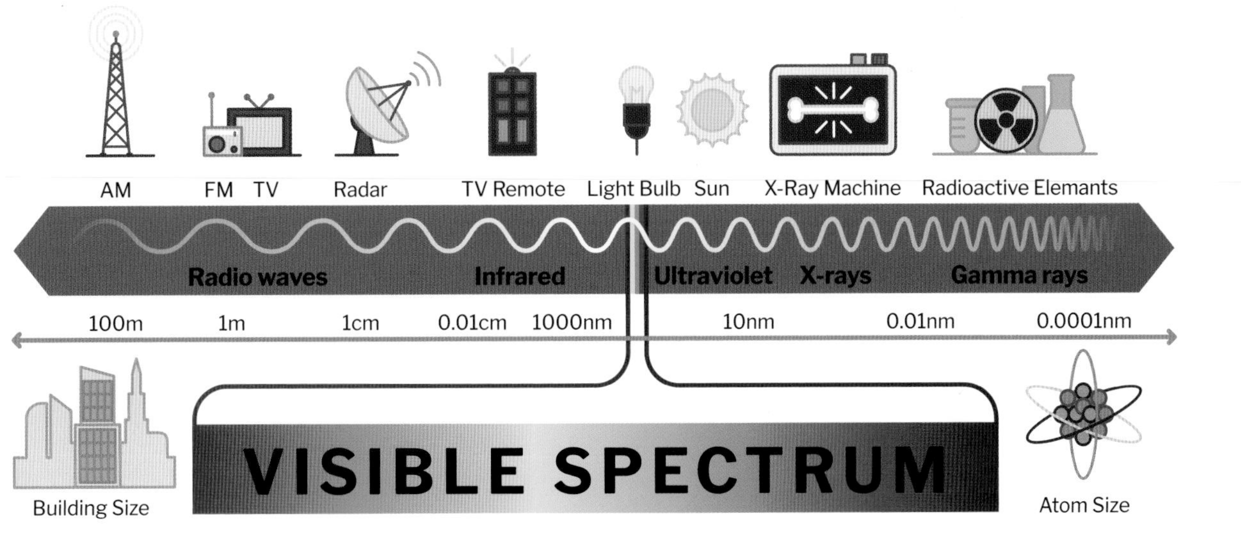

The electromagnetic spectrum ranges from low energy radio waves, which can have wavelengths of thousands of kilometres, to high energy gamma rays with wavelengths on the atomic scale.

20

Perfect harmony

The physics of sound

Sound is matter in motion, transmitting energy in waves. For most people, sound is one of the fundamental ways they learn about what is going on around them. Sound can alert us to danger, allows us to communicate with each other and can entertain us through music.

Greek philosopher Pythagoras is said to have carried out investigations of the sounds made by different-sized, but similarly shaped, objects.

The 6th-century BCE Greek philosopher Pythagoras was one of the first to make a scientific study of the sound of music. He conducted experiments into the relationship between the size of an object and the tone it produced, plucking strings of different lengths and striking vessels filled with varying quantities of liquid to see how the notes changed. He discovered that the intervals between harmonious musical notes always have whole-number ratios and established a mathematical relationship between object and sound.

Pythagoras related the pitch of a plucked string to its length, finding that a string half the length of another will vibrate with a frequency twice that of the longer string and produce a note one octave higher, and that the sounds made by a pair of strings plucked together will sound harmonious if the length of one of the strings is a simple fraction of the other.

Sounds are caused by vibrations, rapid back and forth motions. Clapping

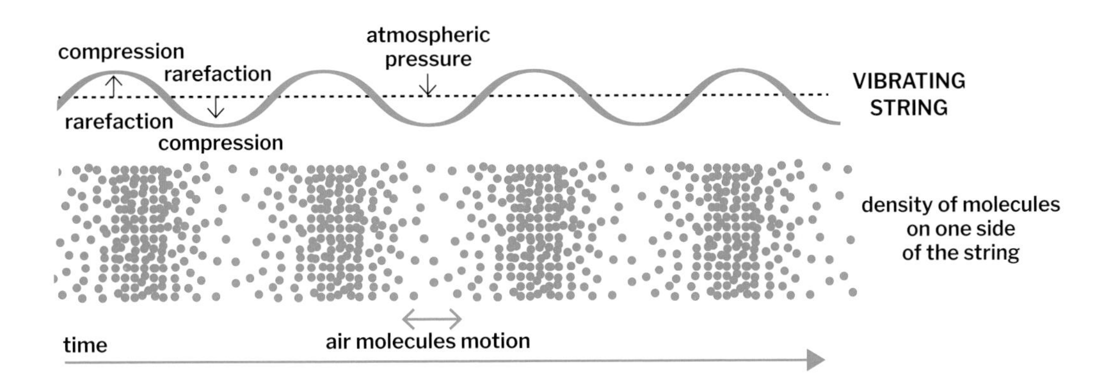

Sound is the result of a succession of compression waves, which are generated by a vibrating object, passing through matter.

your hands or plucking a guitar string produces vibrations. These generate compression waves which travel through the air to our ears where we hear them as sounds. When something vibrates it pushes against the particles next to it. The particles are alternately squeezed together and pulled apart as the vibrating object moves back and forth. The moving particles in turn push against the particles next to them. In this way, a series of compressions (regions of higher pressure) and rarefactions (regions of lower pressure) travels out from the source of the vibrations. The faster the vibrations the higher the frequency of the waves produced and the shorter the wavelength.

The more energetic the vibrations, the higher the amplitude of the waves produced and the louder the sound. The higher or lower the frequency of the vibrations (measured in Hertz, Hz) the higher or lower the pitch of the sound we perceive. The musical note A above the middle C of a piano has a frequency of 440 Hz. This is known as the reference tone – the one other instruments tune to. Doubling the frequency to 880 Hz produces an A that is an octave higher.

A steady pure tone of a single pitch produces a sine wave, a smooth repetitive oscillation. The sounds produced by musical instruments are much more complex. The quality, or timbre, of the sound of a musical instrument results principally from the harmonic content, the number and relative intensities of the upper harmonics present in the sound. The result is a sophisticated waveform with a variety of waves interfering with each other.

A sound can be characterized in terms of the amplitudes of the constituent sine waves it is composed of. This set of numbers is sometimes referred to as the harmonic spectrum of the sound. Once

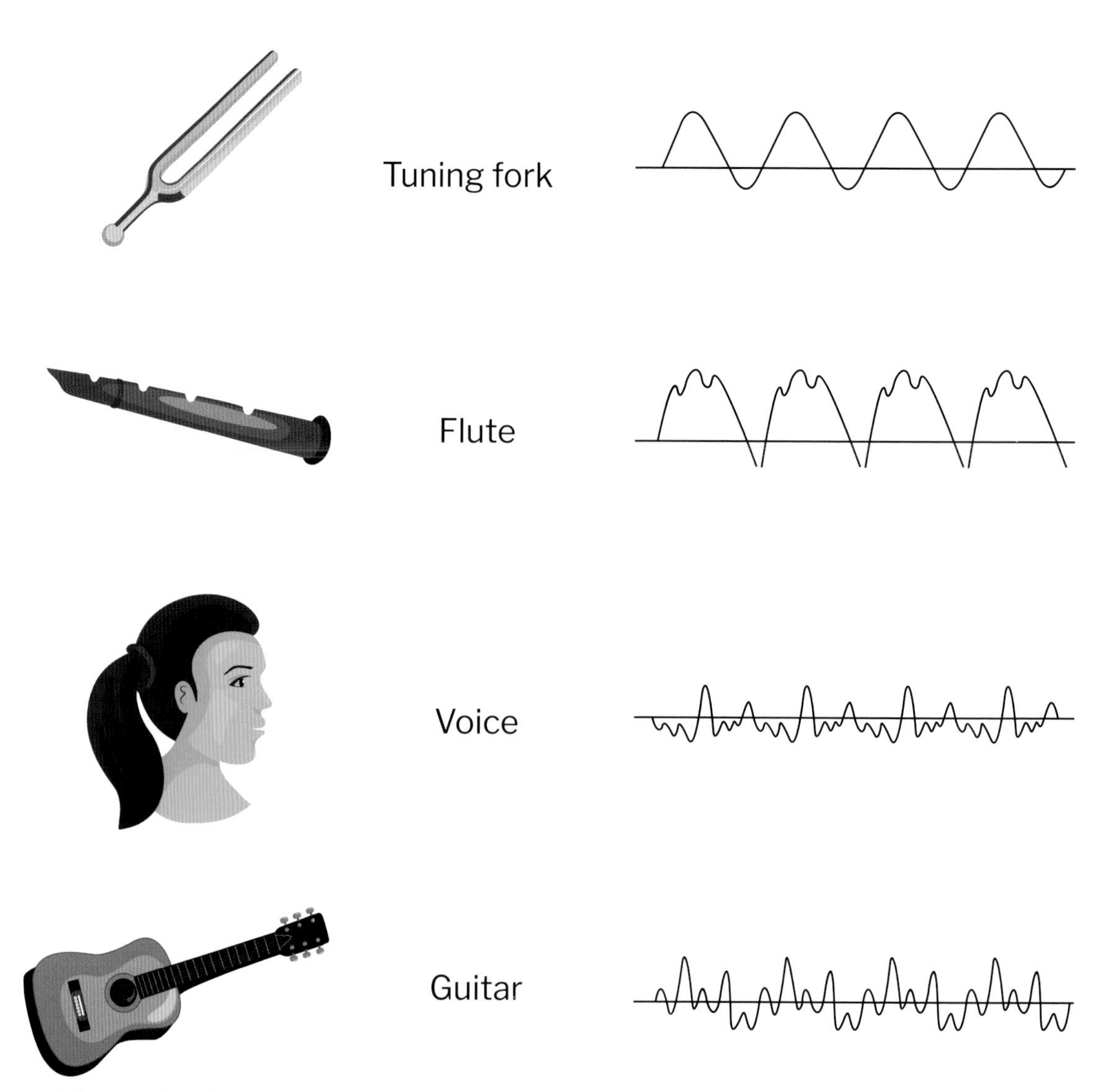

The sound quality, or timbre, of different musical instruments is due to the complex waveforms they produce.

you know the harmonic content it is possible to reverse the process and build up a synthesized version of the original sound using tone generators to create the constituent sine waves. In the 19th century, French scientist Jean-Baptiste Joseph Fourier, while researching the way heat transfers from one place to another, showed that, no matter how complex the waveform, it could be broken down into its constituent sine waves, a process now called Fourier analysis. A variety of modern technologies such as noise-cancelling headphones and speech recognition software rely on Fourier analysis.

21

Chasing rainbows

Light reflected, refracted and focused

For centuries people have sought to explain the various phenomena associated with light. The way light behaves has always been a source of fascination and wonder. The bright light of a full moon, the colours of a rainbow or our own reflection in a mirror quite literally catch the eye.

Different types of materials and surfaces will affect light in different ways. Two of the fundamental characteristics of the way light behaves are reflection and refraction.

When light strikes a surface, or a boundary, some of it bounces back away from the surface. If light wasn't reflected in this way, we wouldn't see anything that wasn't luminous. The reflection from a smooth, flat surface, such as a mirror, is called specular reflection; all the light is reflected evenly, producing a clear image of whatever is being reflected. Light bouncing back from a rough surface is scattered in many directions. This is diffuse reflection, or diffuse scattering.

REFRACTION

Refraction is the change in direction of a wave when it crosses the boundary

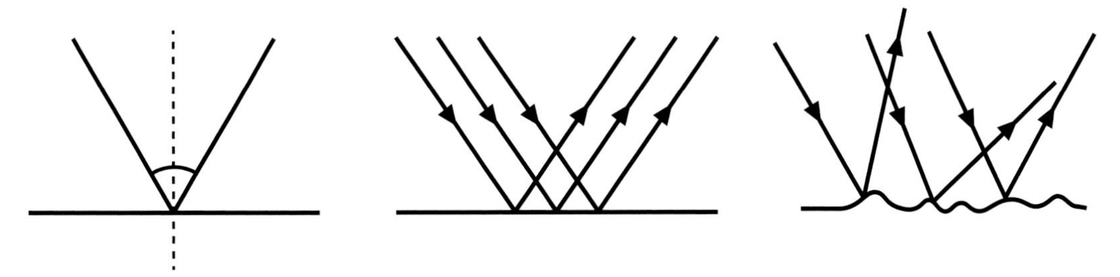

We can only see non-luminous objects because they reflect light. A smooth surface reflects light evenly, a specular reflection, while an uneven surface produces a scattered or diffuse reflection.

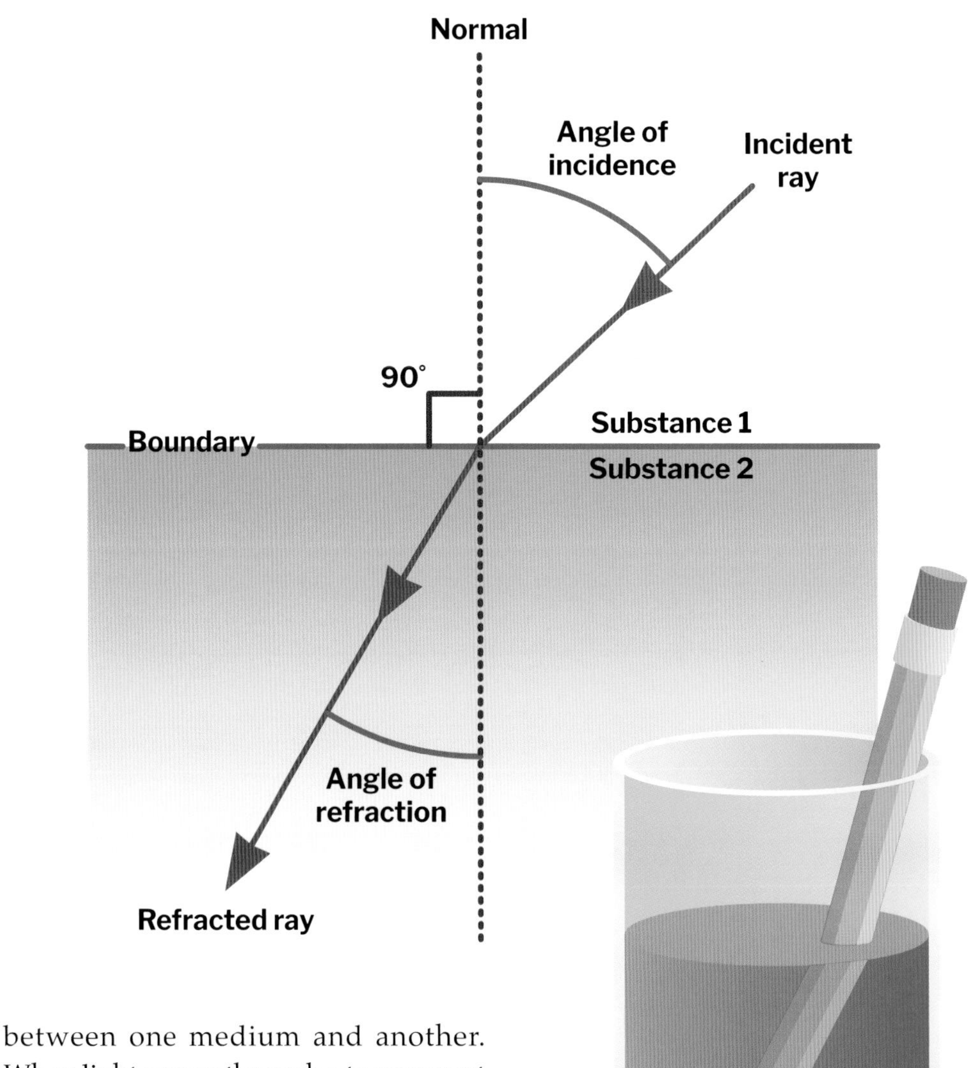

between one medium and another. When light passes through a transparent medium, such as glass or water, it is refracted – its direction of travel is altered. If light enters a denser medium it is refracted towards the normal; if it enters a less dense medium it is refracted away from the normal. At the same time, some of the light is reflected from the surface of the medium or from the boundary between two different media.

Light waves change direction when they pass from one medium to another of different density, a phenomenon known as refraction.

Snell's law

The refractive index is a measure of how far light bends when it enters a different medium. It equals the speed of light in a vacuum divided by the speed of light in the medium. The refractive index of glass is about 1.5. The law relating the angle of incidence (i), the angle of refraction (r) and the refractive index (n), known as Snell's law, was formulated in 1621 by Willebrord Snell. It states that: the refractive index (n) = sin i/sin r

Willebrord Snell.

This change in direction is the result of a change in the wave's speed. When a wave crosses the boundary between different media its frequency stays the same but its wavelength and wave speed change. The change in wavelength is proportional to the change in wave speed. In general, the denser the medium the slower the wave speed.

Refraction doesn't occur if the wave crosses the boundary between media at a 90-degree angle (the normal). If the wave crosses the boundary at an angle away from the normal, one part of the wavefront will change speed before the other, so for the wavefront to stay connected, the wave has to change direction. If the wave speed gets faster, it bends away from the normal. If it gets slower, it bends towards the normal.

TOTAL INTERNAL REFLECTION

As the angle of incidence increases, so too does the angle of refraction. When the angle of incidence reaches the critical angle, the angle of refraction is 90 degrees and the ray will travel along the boundary between the two media. The critical angle is different for different materials; for glass it is around 42 degrees. The critical angle is always in the denser medium. If the angle of incidence is increased beyond the critical angle the light will be totally internally reflected. Total internal reflection only occurs when the light travels from a denser medium to a less dense medium. Prisms shaped to produce total internal reflection are used in optical instruments such as binoculars as they produce a clearer, brighter image than a mirror.

SIGHT CORRECTIONS

Refraction can be put to practical use by using lenses to redirect and focus beams of light. A simple refracting telescope consists of two converging lenses. The objective lens produces a real, inverted image and the eyepiece lens is used to magnify this. The magnifying power of the telescope is calculated by dividing the focal length of the objective lens by the focal length of the eyepiece lens.

A person with near, or short, sight can focus on near objects but not on distant objects. Someone with far, or long, sight has the opposite problem, seeing distant objects clearly but not close ones. Both conditions can be helped using lenses to change the focus of the light entering the eye. Convex, or converging, lenses bulge out at the middle and focus light rays at a point. Concave, or diverging, lenses curve inwards at the middle and cause light rays to spread out.

The light from the sun is made up of a spectrum of different wavelengths of light. Refraction can split the spectrum into its component colours. The shorter the wavelength of the light, the more it is refracted. Violet light, which has the shortest wavelength, is refracted more than the shorter wavelength red. The result is that the light refracted through a prism is spread out, or dispersed, to produce the familiar spectrum.

The beauty of a rainbow is the result of a combination of reflection and refraction. If sunlight strikes a raindrop at just the right angle the light is refracted as it enters the drop, reflected inside it and then refracted again as it leaves it. It is only possible to see a rainbow in a very specific direction. For rainbows to be visible in the sky, the sun must be behind the observer. When the observer looks away from the sun and towards the rain, there is an arc of the sky in which the angle between the sun, the raindrops and the observer is 42° – here red light reflected from the raindrops is visible. Where the angle is 41°, green light is reflected and at 40° violet light is visible. As the angle between the sun, the raindrops and the observer is always the same the curvature of the rainbow's arc will always be the same, no matter how near or far the raindrops are. The reason some rainbows seem bigger than others is that they are basically an optical illusion and do not exist at any specific point in the sky. Your brain attempts to judge the size of the rainbow arc as it compares it to other objects in view. For example, a rainbow over a background of distant hills will appear to be of an impressive size, but a rainbow formed by a garden sprinkler will appear much more modest.

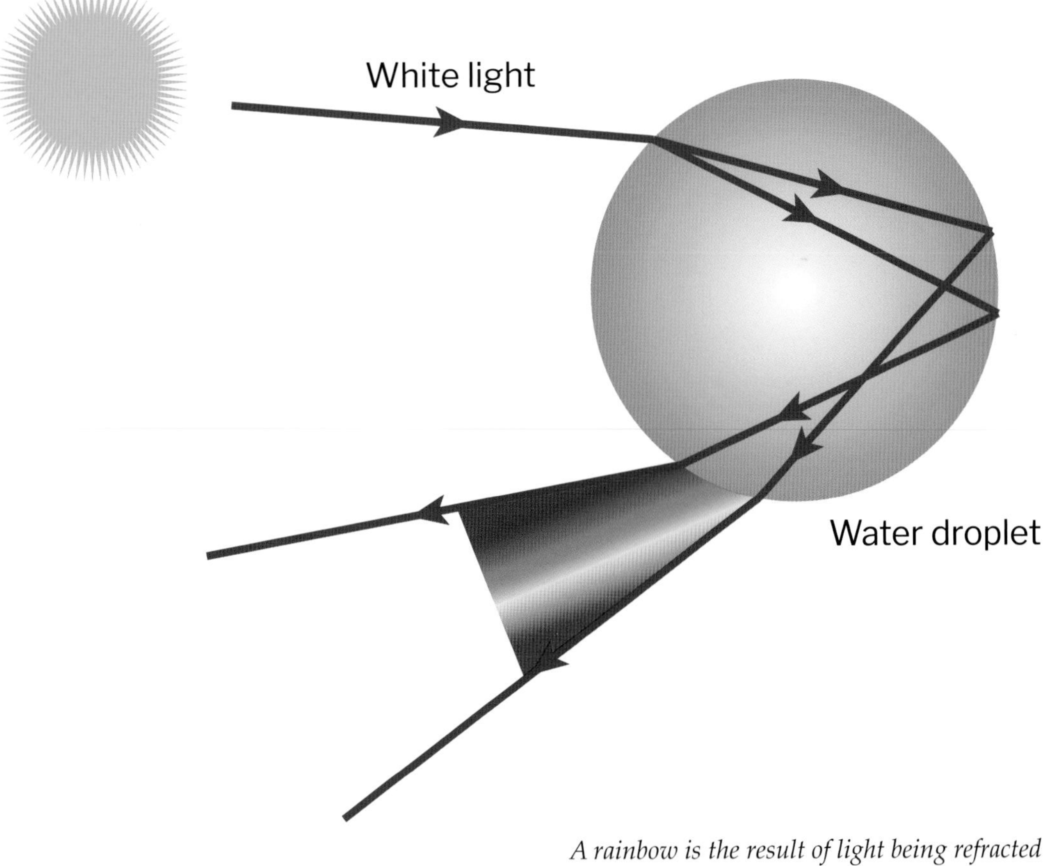

A rainbow is the result of light being refracted and reflected through droplets of water.

22

Interfering with Isaac

The wave nature of light

Ideas about the nature of light have occupied scientists and philosophers for hundreds of years. Not only about the way it behaves but what it actually is. For centuries, light and colour were believed to be two distinct phenomena. Light was thought to be a carrier of colour rather than the source of colour itself.

Philosopher René Descartes suggested that colour might be caused by the rotation of the particles that formed a beam of light and that therefore colour was a property of the light itself. Isaac Newton's experiments with a prism showing that white light could be split into a rainbow of colour and that each colour is refracted by a different degree offered conclusive proof that colour was indeed a property of light.

Isaac Newton's investigations of light revealed that colour was a property intrinsic to light itself rather than a property of matter.

Newton, like Descartes, was a firm believer that light was a stream of minute particles and that the refraction of light through the prism to produce a spectrum was caused by these particles being deflected to a greater or lesser degree according to their size as they passed through. He also attempted to account for the partial reflection of light from transparent materials by suggesting that the light particles were sometimes liable to be reflected rather than transmitted.

Newton noticed that when he placed

a convex lens on a flat glass surface, so there was a thin layer of air between the two, a series of concentric light- and dark-coloured bands were visible. Newton surmised that this was due to the particles of light being set vibrating back and forth between the glass surfaces. The phenomenon, which came to be called Newton's rings, is actually caused by light waves reflected from the top and bottom surfaces of the air film interfering with each other but Newton was reluctant to accept that light had wavelike properties.

Seventeenth-century Italian physicist Francesco Maria Grimaldi examined the transmission of light through a small hole. He observed that a small amount of light could be seen in regions that would have been expected to be in shadow had the light followed the straight-line course that a stream of particles would have been expected to take, noting that the light pattern formed complicated coloured bands. He named this phenomenon diffraction and speculated that it suggested light might travel in a wavelike manner.

Newton's contemporary, scientist Robert Hooke, was one of the first to make a link between the diffraction of light and the similar behaviour of water waves. This, he suggested, indicated that light was also a wave. Further, he believed that light, like water, was a transverse wave, one in which the direction of motion, or propagation is at right angles to the direction of the wave vibration, an assumption that later research would prove to be correct.

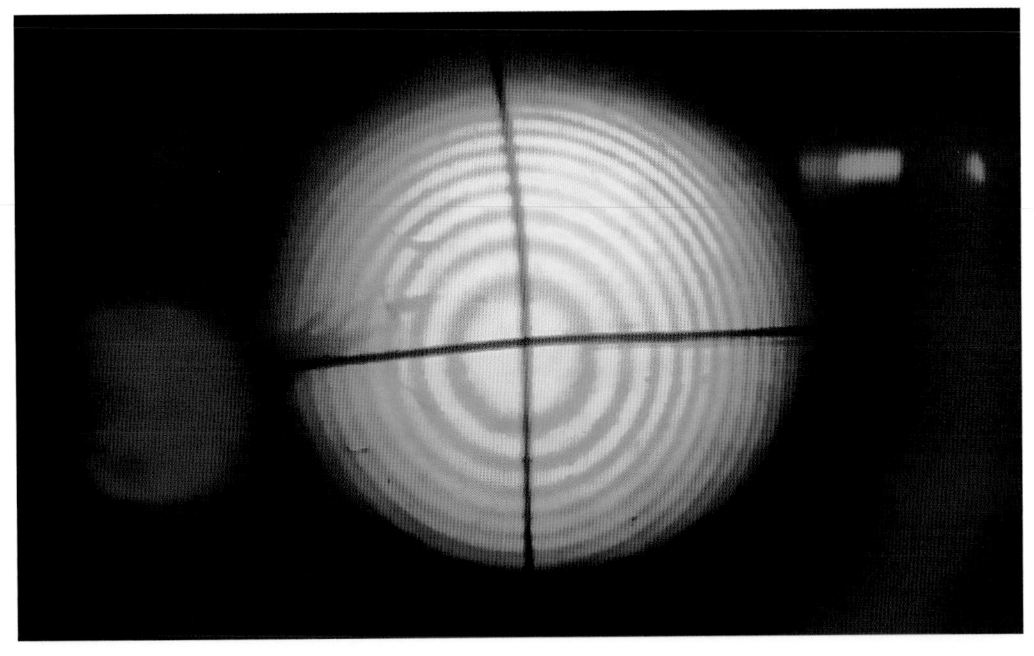

Newton's rings seen through a microscope.

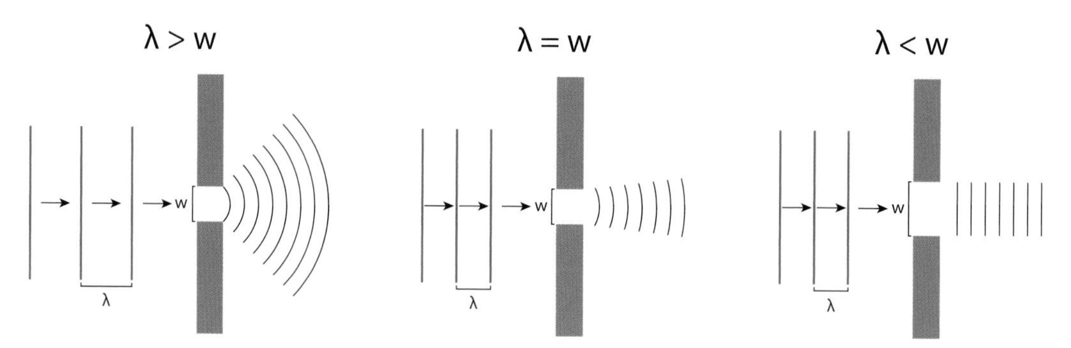

When a wave passes through a narrow aperture it spreads out, a phenomenon known as diffraction. The smaller the aperture in relation to the wavelength, the greater the diffraction.

Christiaan Huygens believed in the wavelike nature of light, publishing his ideas in his *Treatise on Light* in 1690. As was well understood, a wave needs a medium to travel through, as, for example, sound waves travelling through air, so Huygens proposed that light travelled through an invisible, but all-pervading medium called ether. He theorized that every point on a wavefront could be considered as a source of secondary spherical wavelets which spread out in the direction of travel of the light wave at the speed of light. It was an elegant theory and explained most of the observed phenomena of light such as reflection, refraction and diffraction. Unfortunately, such was the influence of Newton on physics, Huygens, Hooke and Grimaldi were more or less ignored and the particle theory of light remained the generally accepted one.

'PHENOMENON' YOUNG

Thomas Young was a man of such formidable intellect that his fellow students at Cambridge University dubbed him 'Phenomenon'. Between 1801 and 1803 he delivered a series of lectures to the Royal Society in London. In his talk of 1801, he put forward his theory of three-colour vision to explain how the eye detected colours, an idea that was finally confirmed in the 1950s.

Also in 1801, Young described interference, the pattern formed when waves meet. Rather than bouncing apart waves appear to pass straight through each other, combining where they intersect. If the peak of one wave meets the peak of another they add together to make a higher peak; two troughs make a deeper trough and a trough and a peak cancel each other out. The result was an interference pattern that showed where the waves were adding and cancelling. Young hypothesized that if light was wavelike, rather than a stream of particles, then it should produce interference patterns similar to the ripples on a pond.

A small hole in a window blind provided a point source of illumination which he directed on to a screen through two pinholes placed close together on a board. If Newton was right, the streams of light particles would produce two discrete points of light on the screen where they travelled through the pinholes. Instead Young saw a series of curved, coloured bands separated by dark lines, exactly the interference pattern that would be expected if light were a wave. Young called these bands, which appeared to be convincing proof of the wave nature of light, interference fringes. Young's experiment was convincing proof of the wave nature of light. He theorized that if the wavelength of light was sufficiently short it would explain why it appeared to travel in straight lines, as if it were a stream of particles.

The question still remained: What actually was light? As science began to explore the quantum realm in the 20th century, Thomas Young's two-slit experiment would resurface as one of the most profound in all of physics.

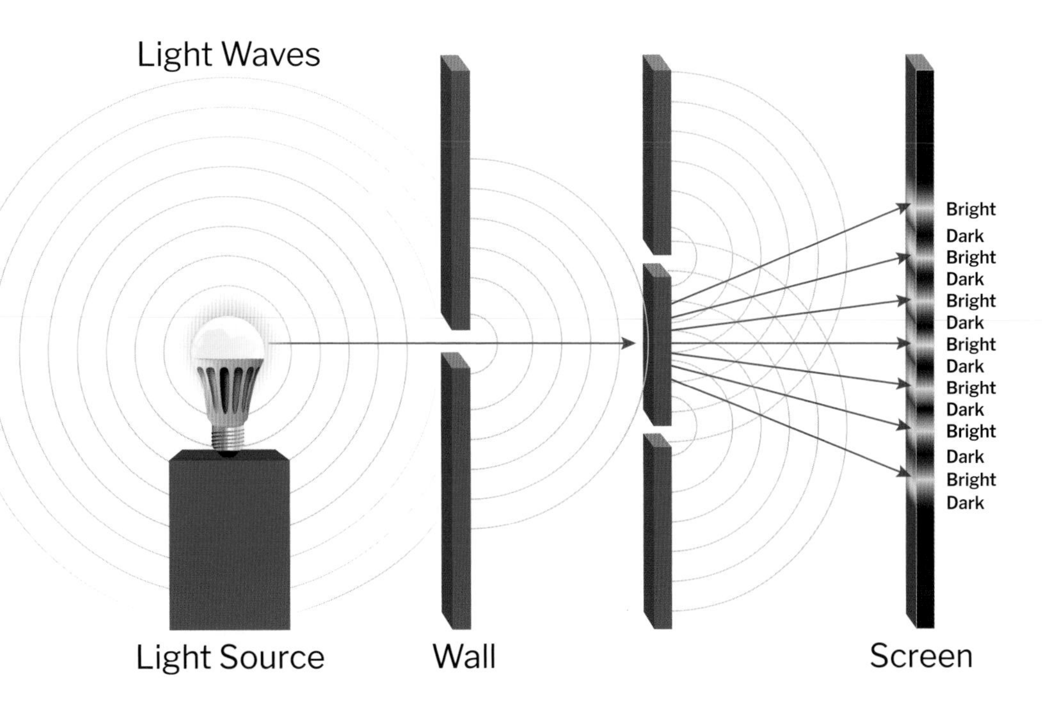

Thomas Young's demonstration of interference between two sources of diffracted light proved beyond doubt that light was a wave.

23

Light from the atom

Luminescence and spectral emissions

It has been known for centuries that certain substances have the ability to generate light but without producing any noticeable heat. It wasn't until the 19th century, however, that scientists began to study this curious phenomenon. In time, their investigations would lead to the heart of the atom.

One of the first to investigate fluorescence was clergyman Edward Clarke who, in 1819, described how the mineral fluorspar sometimes glows. It was Irish physicist George Stokes who showed a few decades later that the glow was caused by exposure to ultraviolet light. Stokes coined the term 'fluorescence' to describe this behaviour and believed that the shorter wavelength and invisible

Fluorescent materials, such as the mineral fluorite, absorb ultraviolet light and release it at a different wavelength, changing their colour.

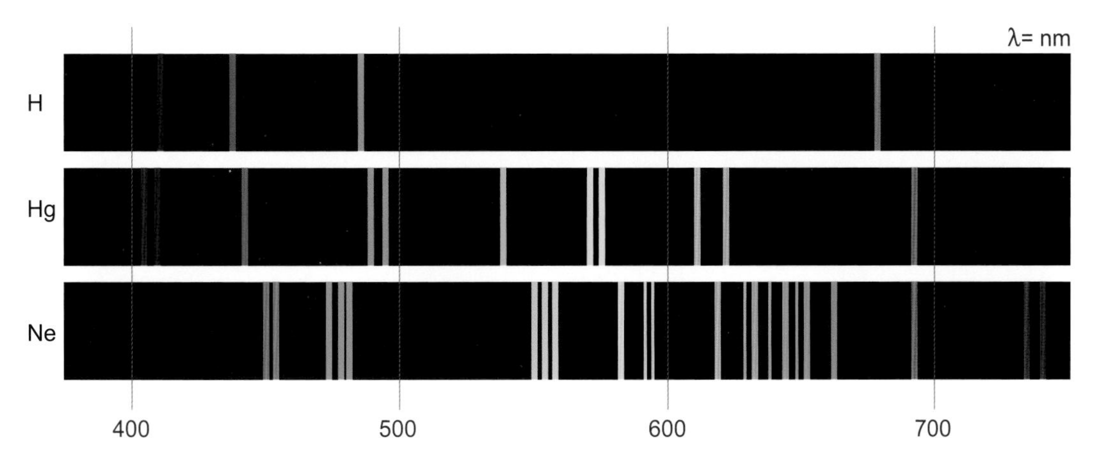

When heated, different elements emit light at characteristic wavelengths, producing unique patterns of emission lines.

to the human eye ultraviolet light was being transformed by the fluorspar into longer wavelength visible blue light.

In the late 1850s German scientists Robert Bunsen and Gustav Kirchhoff began investigating why some atoms produce light. They discovered that the light produced by different elements heated to incandescence was of very specific wavelengths, producing just a few bright lines of colour, called emission lines, rather than a continuous spectrum. The precise nature of the colours produced was unique to each element. Bunsen and Kirchhoff identified two previously unknown elements, caesium

Luminescence

There are three main forms of luminescence: fluorescence, phosphorescence and chemiluminescence. Fluorescence and phosphorescence are both forms of photoluminescence. In photoluminescence, the glow is triggered by light, in chemiluminescence, the glow is the result of a chemical reaction. Both fluorescence and phosphorescence are the result of a substance absorbing light and emitting light of a longer wavelength and therefore lower energy. In fluorescence, the emission is immediate and only visible when the triggering light source is on; phosphorescent material can store the absorbed light energy and release it later, resulting in an afterglow that can last anywhere from a few seconds to hours depending on the material.

and rubidium, just from their emission lines.

BOHR'S BREAKTHROUGH

An important breakthrough in understanding how the emission lines were linked to the properties of atoms came in 1885 when Swiss mathematician Johann Balmer discovered that the emission lines of hydrogen could be predicted by a formula, which came to be called the Balmer series. The accuracy of the formula was proved when the predicted hydrogen wavelengths were detected in the spectrum of the sun and it was soon being adapted to predict emission lines from other elements too.

At this time atoms were believed to be solid particles, so it was hard to explain how the emission lines were produced. J.J. Thomson's discovery of the electron in 1897 and Ernest Rutherford's discovery of the atomic nucleus in 1909 revealed that the structure of the atom was more complex than had been believed. Danish scientist Niels Bohr linked his ideas about the structure of the atom to the emission lines formula. Inspired by Max Planck's revolutionary theory that radiation was emitted in discrete parcels called quanta, Bohr suggested that electrons orbited at fixed distances from the nucleus, with low energy orbits closer to the nucleus and higher energy orbits further out. If an electron received an energy boost by absorbing a quantum of light it could jump from a lower orbit to a higher one. A quantum that was energetic enough could eject the electron altogether. When the electron fell back to the lower orbit it would emit a pulse, or quantum, of light. These steps up and down energy levels are what we now call quantum leaps.

The Bohr model successfully linked spectral lines with the structure of the atom.

He applied his new model to the hydrogen atom and showed how it explained the emission lines in the Balmer series. He also explained the emission lines found in superhot stars by determining that these were associated with electrons jumping between orbits in atoms of ionized helium. Although later research into quantum mechanics would show that the Bohr atomic model was an oversimplification, it was still an important step forward.

24

The dawn of the quantum

Planck's quanta and the photoelectric effect

Until the beginning of the 20th century physics was largely believed to be a deterministic science. Experiments had uncovered laws that allowed physicists to correctly predict how the universe worked, how it became the way it was, and how it would change over time. The physics that was about to unfold would unsettle that viewpoint and show that at its heart reality was strangely indeterminate.

Nineteenth-century theory predicted that the electromagnetic radiation emitted from a black body, an object that is a perfect absorber and reflector of radiation, should become infinite for shorter and shorter wavelengths. Everyday observation showed that this prediction, which was dubbed the ultraviolet catastrophe as it concerned short wavelength radiation beyond the ultraviolet, was obviously wrong, but no one could explain why.

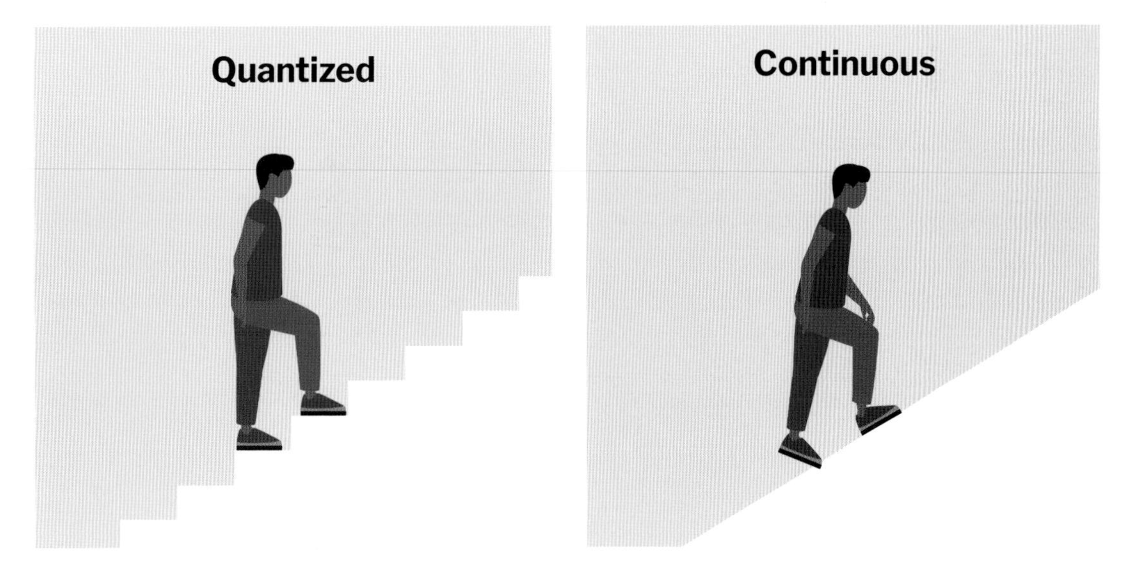

Max Planck theorized that energy, rather than being continuously variable, came in discrete packets, or quanta.

Planck's constant

Planck suggested that the energy of a quantum was related to its frequency using the simple formula E=ħv where E equals energy, v equals frequency and ħ equals a value known as Planck's constant. The energy of a quantum can be calculated by multiplying its frequency by Planck's constant, which is 6.62607015 x 10-34 J/s (joule-seconds).

Max Planck opened the door to a revolution in our understanding of physics.

On 14 December 1900, Max Planck made a presentation to the Deutsche Physikalische Gesellschaft (the German Physical Society) in Berlin in which he submitted a radical solution to the ultraviolet catastrophe problem. Planck made the revolutionary suggestion that the energy emissions of the black body, rather than being a continuously variable quantity in fact came in discrete packets, which he called quanta (singular quantum) from a Latin word meaning 'how much'.

Energy could only be emitted in whole quanta, the size of which was proportional to the frequency of vibration. Although in theory there were an infinite number of higher frequencies, it took increasingly large amounts of energy to release quanta at the higher levels. For example, the frequency of a quantum of violet light is twice that of a quantum of red light and therefore has twice the energy. This proportionality also explains why a black body doesn't emit energy equally across the electromagnetic spectrum.

There was no doubt that Planck's solution worked – the results of experiments agreed with the predictions made by his theory. Nonetheless, Planck wasn't entirely happy with his explanation. He saw his quanta as more of a mathematical 'fix' for a difficult problem rather than something that had any basis in reality. He spent years looking for a way to disprove his own theory, which he admitted he had introduced as 'an act of desperation'. The unforeseen consequences of Planck's desperate act were to lead to a radical overhaul of the world of physics. When Albert Einstein heard about Planck's theory he commented: 'It was as if the ground had been pulled out from under us.'

THE PHOTOELECTRIC EFFECT

In 1887, German physicist Heinrich Hertz had discovered that certain types of metal would emit electrons when a beam of light was directed at them. This was known as the 'photoelectric effect' and is the phenomenon harnessed by devices such as solar cells to generate electricity from light.

At first it was assumed that the electric field part of the electromagnetic wave provided the energy the electrons needed to break free from the metal. If this theory was correct, then the brighter the light the higher energy the emitted electrons should be, but experiment showed that the energy of the electrons released depended not on the intensity of the light but on its frequency. Shifting the beam to higher frequencies, from blue to violet and beyond, produces higher energy electrons; a low frequency red light, even if it is blindingly bright, produced no electrons. It was as if fast-moving ripples could readily move the pebbles on a beach but a slow-moving wave, no matter how big, left them untouched. In addition, if the electrons were going to jump at all, they jumped

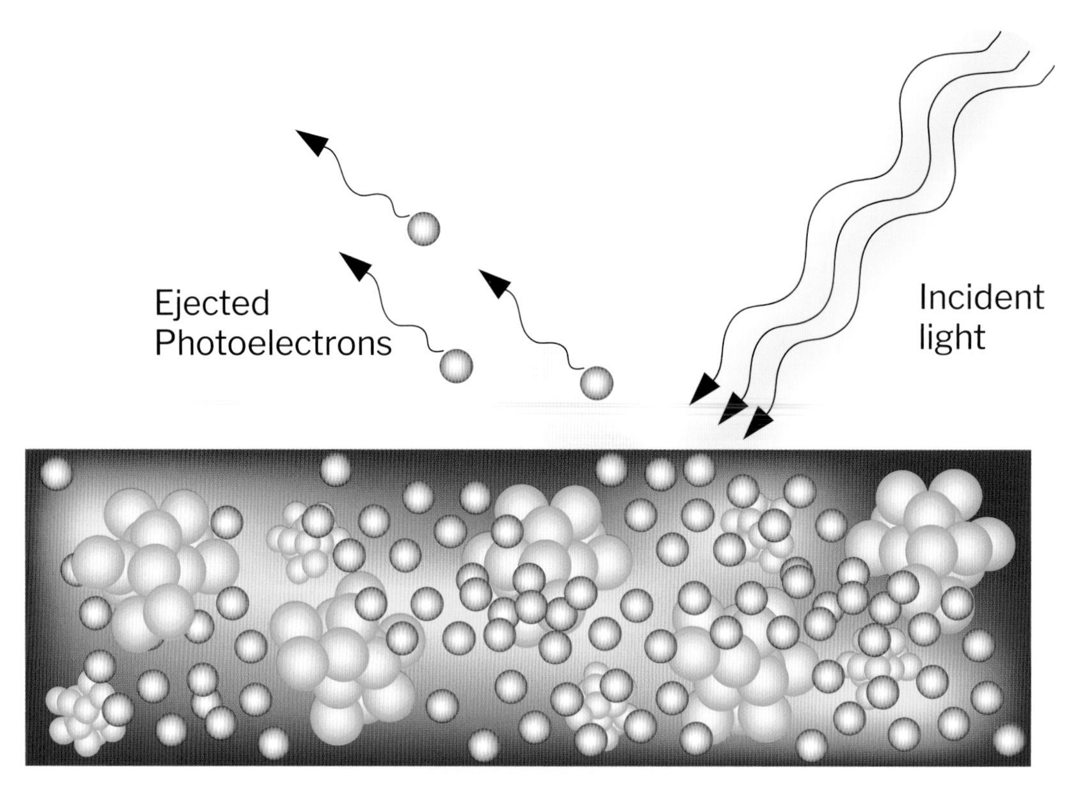

The emission of electrons by some metals when they are struck by light is known as the photoelectric effect. The higher the frequency of the light, the higher the energy of the electron emitted.

right away – there was no build up of energy involved. The wave theory of light could make no sense of these findings.

In 1904, Einstein wrote to a friend that he had discovered 'in a most simple way the relation between the size of elementary quanta ... and the wavelengths of radiation'. In March 1905, he published a paper in the journal *Annals of Physics* that took Planck's quanta and married them to the photoelectric effect. This was the work that would eventually win him the Nobel Prize in 1921. What Einstein did was to look at the differences between particle theories and wave theories. He compared the formulae that described the way the particles of a gas behave as it changes volume with those describing how waves of radiation spread through space. He found that both obeyed the same rules and the mathematics underpinning both phenomena was the same. He wrote that light 'consists of a finite number of energy quanta ... which can be produced and absorbed only as complete units.'

Einstein used these insights to calculate the energy of a light quantum and found that his results agreed with Planck's. As Planck had established, the energy of a quantum was determined by its frequency. A single quantum transferred its energy to an electron – the higher the energy of the quantum, the more likely it was to cause the electron to be emitted. High-energy blue quanta had the heft to punch out electrons, low-energy red quanta simply couldn't. In this way Einstein showed that the existence of light quanta explained the photoelectric effect.

Whereas previously Planck had considered the quantum to be little more than a mathematical contrivance, Einstein was now suggesting that it was an actual physical reality. It was a suggestion that didn't go down well with other physicists who were hesitant to give up the idea that light was a wave, and not a stream of particles. Planck even suggested that Einstein 'may have gone overboard in his speculations'. In 1915, a sceptical Robert Millikan performed experiments on the photoelectric effect that were aimed at disproving Einstein's assertion, but instead ended up producing results that were entirely in line with Einstein's predictions. Nonetheless, Millikan still persisted in referring to Einstein's 'reckless hypothesis'. In 1923, American physicist Arthur Compton carried out experiments on the scattering of x-rays from electrons. The small change in the frequency of the x-rays that resulted became known as the Compton effect and provided compelling evidence that the x-rays and the electrons were behaving as streams of particles when they collided. He declared that his experiments left him with little doubt that x-ray scattering was a quantum phenomenon.

From here the quantum theory gained growing acceptance. Experiment had verified Einstein's explanation of the photoelectric effect and light, almost in defiance of accepted knowledge, acted as if it were a stream of particles. Yet indisputably experiment after experiment in reflection, refraction and interference had shown that it behaved like a wave. What could possibly be going on?

25

How can it be like that?

Unravelling the nature of light

Arguments on the nature of light stretch back millennia. The ancient Greek philosopher Aristotle saw it as a wave passing through an invisible ether. The Roman philosopher Lucretius thought it a stream of minute particles shooting out from the sun. Einstein's found convincing evidence for the particle nature of light. So was it a wave or a particle?

THE PERPLEXING DOUBLE SLIT

At the beginning of the 19th century Thomas Young had convincingly demonstrated that light was a wave by showing how it formed interference patterns when it passed through twin slits. In the early 1960s physicist Richard Feynman returned to the twin slits in a thought experiment that remains resolutely mind-boggling.

Feynman described an experiment in which he imagined what would happen if just one photon or electron at a time was directed towards a detector through twin slits that could be opened or closed. Common sense would suggest that the photons would travel as particles, arrive as particles, and be detected on the screen as individual dots. There should be two bright areas when both slits were open or just one if a single slit was open. In neither case would you expect to see an

What is a photon?

Einstein referred to the particles of light he described in his photoelectric effect paper as lichtquant (light quanta). The first person to use the word 'photon' to describe a unit of light was chemist Gilbert Lewis in a 1926 letter to the journal *Nature*, although he thought of it as a carrier of radiant energy rather than as a particle of light. A photon is the smallest discrete amount or quantum of electromagnetic radiation. Radio, microwaves, visible light and x-rays are all carried by photons.

Richard Feynman described an experiment in which quanta of light, travelling as discrete particles, still interfered as if they were waves.

interference pattern forming. Feynman, however, predicted a different outcome. What happens is that the pattern on the screen does build up, particle by particle, into an interference pattern when both slits are open. Even if subsequent photons are fired off *after* the earlier ones have hit the screen they still somehow 'know' where to go to build up the interference pattern. It is as if each particle travels as a wave, passing through both slits simultaneously, and interferes with itself! No pattern forms if only one slit is open.

How can that be? How does a single

particle travelling through the left-hand slit know whether the right-hand slit is open or closed? Feynman advised against even attempting to answer these questions, declaring it *'absolutely* impossible, to explain in any classical way'.

In 1964 he wrote: 'Do not keep saying to yourself, if you can possibly avoid it, "But how can it be like that?" because you will go down the drain into a blind alley from which nobody has yet escaped. Nobody knows how it can be like that.'

Later experiments confirmed Feynman's predictions. In 1974, researchers at the University of Bologna in Italy demonstrated single-electron double-slit diffraction by passing single electrons through a biprism – an optical device that acts like a double slit – and observed the build-up of a diffraction pattern. A similar experiment was carried out in Japan in 1989. The Italian researchers also performed another experiment in 2012, in which the arrivals of individual electrons from a double slit were recorded one at a time.

CAN A PARTICLE BE A WAVE?

French physicist Louis de Broglie put forward a theory in 1924, in which he proposed that all matter and energy, not just light, has the characteristics of both particles and waves. Believing intuitively in Einstein's quantum theory of light and in the symmetry of nature, de Broglie asked, if a wave can behave

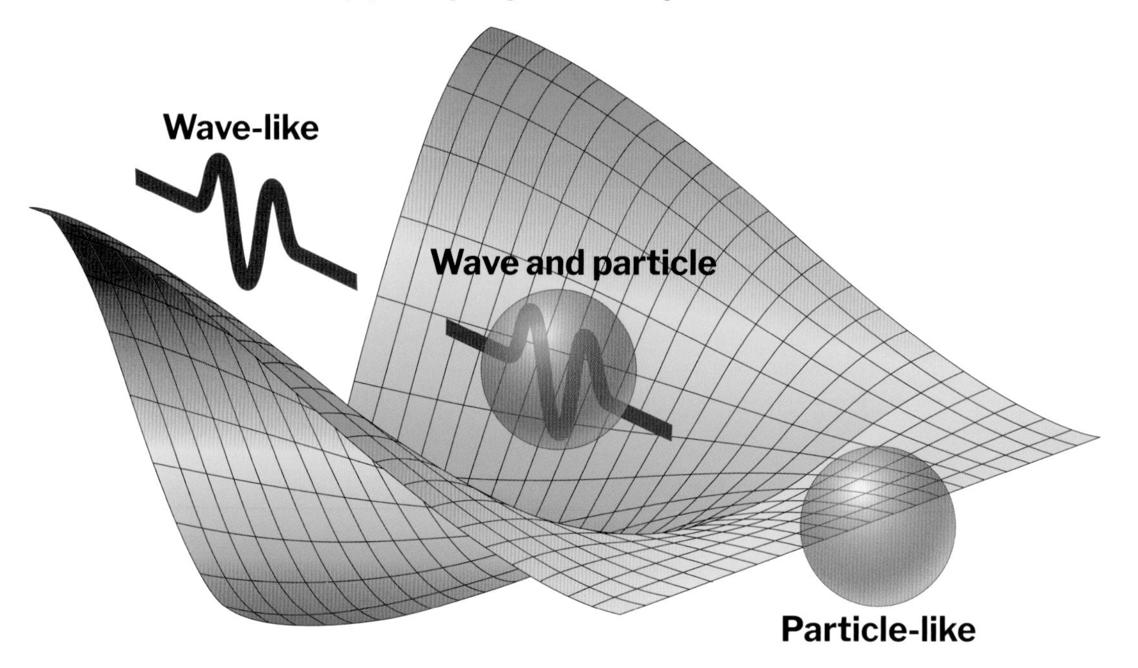

Is light a wave, is it a particle, or something else entirely? Richard Feynman called such speculation a 'blind alley'.

Louis de Broglie.

like a particle, then why can't a particle, such as an electron, also behave like a wave? Einstein's famous $E=mc^2$ equation (see page 162) relates mass to energy and Einstein and Planck had related energy to the frequency of waves so, de Broglie reasoned, combining the two suggested that mass should have a wavelike form as well.

De Broglie came up with the concept of a matter wave, suggesting that any moving object has an associated wave. The kinetic energy of a particle is proportional to its frequency and the speed of the particle is inversely proportional to its wavelength – faster particles have shorter wavelengths. Einstein supported de Broglie's idea as it seemed a natural continuation of his own theories. De Broglie's thesis was experimentally verified in 1927 when researchers in the UK and the United States demonstrated that a narrow electron beam directed through a thin crystal of nickel formed a diffraction pattern as it passed through the crystal lattice.

Perplexing as it might seem, all the experimental evidence suggests that both the wave theory and the particle theory of light are correct. Whether light acts as a wave or as a particle seems to depend on what observations we are making and how it is being measured. We have no single model that can describe light in all its aspects. What this means in terms of the actual physical nature of light is something that no one can answer satisfactorily.

Einstein himself struggled without success to solve the dual nature paradox of light. Towards the end of his life, in 1951, Einstein wrote that 'fifty years of conscious brooding have brought me no closer to the answer to the question, "What are light quanta?" Of course, today every rascal thinks he knows the answer, but he is deluding himself.'

26

Leaps and bounds

Quantum numbers

By the 1920s, experiments had begun to show that light could behave like a stream of particles and that electrons could behave like waves. Scientists were already beginning to challenge Niels Bohr's radical new model of the atom.

Atoms give off light at specific wavelengths, each element producing a characteristic set of spectral lines. Bohr had proposed that these lines were related to the energy levels of the orbits occupied by the electrons around the atomic nucleus and that they were produced by electrons absorbing or emitting a photon at the frequency corresponding to the spectral line.

The energy level that an electron can occupy in an atom is denoted by the principle, or primary, quantum number, n – n = 1 denotes the lowest possible orbit, n = 2 the next highest and so on. The larger n is, the farther the electron is from the nucleus, the larger the size of the orbital, and therefore the larger the atom. The first principal shell is also called the ground state, or lowest energy state. There are no atoms with zero energy so n cannot be 0 or negative. Bohr's scheme described the energy levels in the simplest of atoms, hydrogen, in which a single electron orbits a single proton. Later models had to incorporate

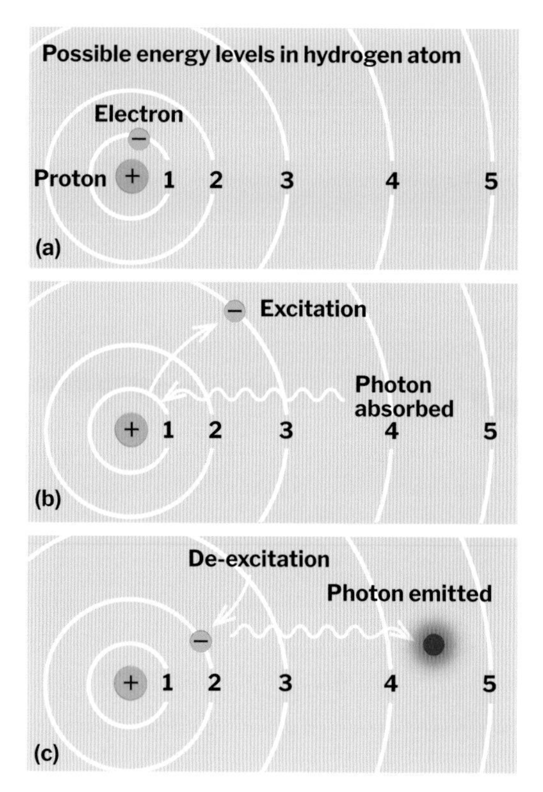

Electrons move up energy levels in an atom by absorbing photons. A photon is emitted when the electron moves down an energy level.

the wavelike properties of electrons to work for larger atoms.

QUANTUM PROPERTY REVEALED

In 1922, Otto Stern and Walther Gerlach carried out an experiment in which they directed a beam of silver atoms through an inhomogeneous, or variable, magnetic field. Stern and Gerlach expected the silver atoms to act like tiny bar magnets and therefore be deflected up or down by the magnetic field, depending on the orientation of each atom's magnetic poles. The random deflections should have resulted in a vertical line appearing on the detector screen. Instead, what appeared were two distinct spots, indicating that the beam had split in two.

Today, the Stern-Gerlach experiment is interpreted as evidence of electron spin. The splitting of the beam is not due to magnetism in the silver atom but to the spin of the atom's outermost unpaired electron. Electrons have a magnetic property because they act like a spinning electric charge and, as classical physics tells us, a moving electric charge generates a magnetic field. Stern and Gerlach's beam split because there are only two directions in which an electron can spin – up or down. (It's important to remember that these are arbitrary labels which don't have the same meaning that they do in 'our' world. Electrons don't actually 'spin' like little balls, and the idea of 'up' or 'down' is meaningless in the quantum realm in the sense of defining a position in space. To describe whether the spin of a particle is up or down, physicists assign a spin quantum number. For electrons

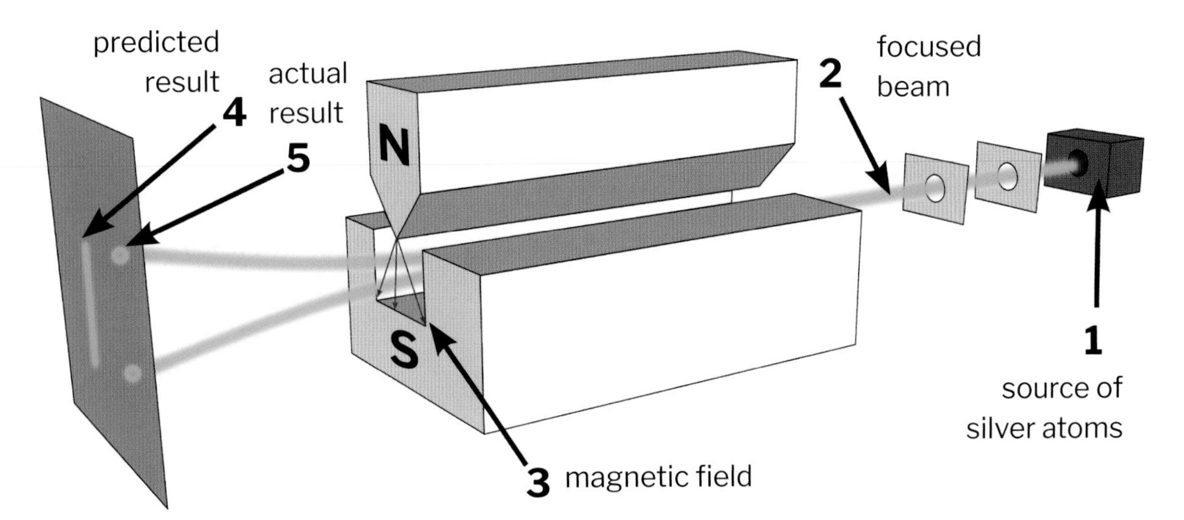

Stern and Gerlach's experiment directed a stream of silver atoms through a powerful magnetic field. The up and down spin properties of the silver atom's outermost electron resulted in the beam splitting in two.

Quantum numbers

A total of four quantum numbers describe the movement and trajectory of each electron within an atom. Each electron has a unique set of quantum numbers:

Principal Quantum Number (n)
Specifies the energy of an electron and the distance of its orbital from the nucleus. If n=1, the electron is in its ground state; if the electron is in the n=2 orbital, it is in an excited state.
Secondary Quantum Number (l)
Measures angular momentum and specifies the shape of an orbital with a particular principal quantum number. The secondary quantum number divides the orbitals into smaller subshells.
Magnetic Quantum Number (ml)
Specifies the orientation in space of an orbital of a given energy (n) and shape (l).
Spin Quantum Number (ms)
Specifies the orientation of the spin axis of an electron and has a value of +½ or -½. An electron can spin in only one of two directions (sometimes called up and down).

this is defined as plus or minus ½. Don't worry about how something can have half a negative spin.) The Stern-Gerlach experiment marked the first time that a particle's quantum properties had been revealed. Further experiments showed that other properties such as angular momentum, a property of rotating objects, and magnetism could be quantized too.

THE PAULI EXCLUSION PRINCIPLE

In 1925 Wolfgang Pauli attempted to explain why atoms were structured the way they were. What decided an electron's energy level and the number of electrons that each energy level, or shell, could hold? He reasoned that each electron had a unique code, described by its four quantum numbers – energy, spin, angular momentum and magnetism. The exclusion principle that he devised stated that no two electrons in an atom could share the same four quantum numbers. No two identical particles could occupy the same state at the same time. Two electrons could occupy the same shell but only if they had opposite spins, for example.

Pauli's principle applies to all the basic building blocks of atoms, protons and neutrons as well as electrons. It explains why we don't fall through the floor, why

Wolfgang Pauli.

two objects can't occupy the same space and why solid matter is rigid. It is one of the most profound principles in physics.

THE SCHRÖDINGER EQUATION

A particle that is behaving like a wave has no fixed position in space. The wave can be thought of as akin to a probability graph mapping out the chances of locating the particle at any particular point in space. In 1926, Austrian physicist Erwin Schrödinger devised an equation that provided a mathematical description of a quantum system. It determined how probability waves, or wavefunctions as they came to be known, are shaped and how they evolve. The wavefunction holds all the information necessary to describe a quantum system and gives all possible locations of a particle at a given time. The quantum numbers that define the properties of a particle arise from solving the Schrödinger equation. Schrödinger tested his equation on experimental observations of the hydrogen atom and found that it predicted its properties with great accuracy.

The Schrödinger equation is as important to the subatomic world of quantum mechanics as Newton's laws of motion were for forces and motion on a large scale. Schrödinger was describing the quantum world in purely mathematical terms, with outcomes that could only be seen in terms of probabilities and not certainties. This was a view of reality that defied any attempt to visualize it in 'real world' terms.

One of quantum mechanics' biggest shortcomings was its failure to take Einstein's relativity theories into account. British physicist Paul Dirac attempted to unite the two concepts and in 1928, he succeeded in marrying Schrödinger's equation to Einstein's famous $E=mc^2$ equation, producing an equation that was consistent with both special relativity and quantum mechanics in its description of electrons and other particles (see page 144). The Dirac equation, which viewed electrons as excitations of an electron field, in the same way that photons could be seen as excitations on the electromagnetic field, became one of the foundations of quantum field theory.

27

Here, there and everywhere

Heisenberg's uncertainty and Schrödinger's cat

At the beginning of the 1920s scientists were challenging Niels Bohr's model of the atom. Experiments had established that electrons could act like waves and form an interference pattern just as light did, and that light sometimes appeared to act as if it were a particle. How were these contradictions to be resolved?

In June 1925, German physicist Werner Heisenberg made a breakthrough. He asked what it actually means to define the position of a particle. In order to know where something is, said Heisenberg, we have to interact with it. To discover the position of an electron we have to bounce a photon off it – the higher the frequency

Werner Heisenberg's uncertainty principle puts a limit on what it is possible for us to know about the quantum realm.

of the photon the more accurately we can position the electron. Unfortunately, the higher the frequency of the photon, the more energy it carries and so the more it will change the trajectory of the electron. We might know where the electron is at that precise instant, but we will have no idea where it is going. If we could precisely measure the momentum of the electron its location would be completely uncertain and vice versa. Heisenberg showed that this principle of uncertainty was a fundamental property of the universe that put a hard limit on what it was possible for us to know at any time.

PROBABILITY WAVES

The wave nature of matter meant thinking in terms of probabilities rather than certainties. The best that we can

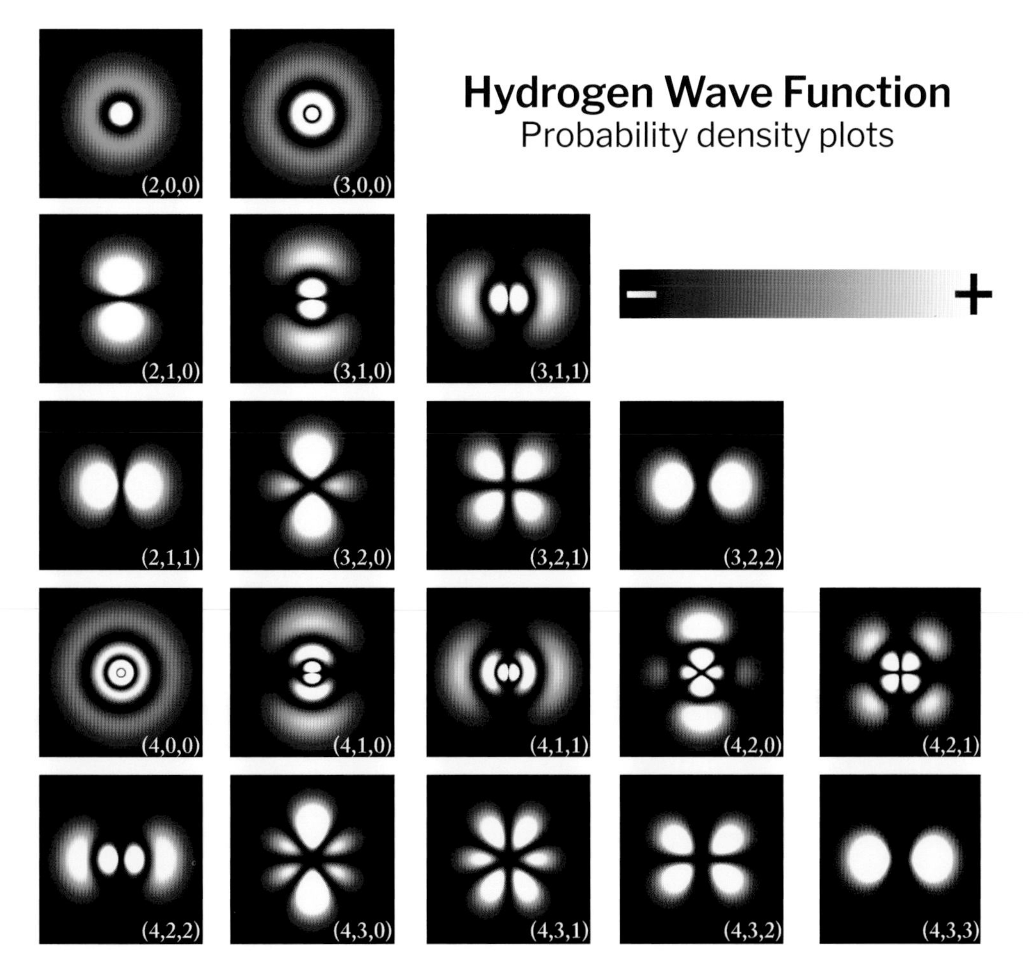

Maps of the electron orbitals of hydrogen based on the energy and momentum of the electron. Rather than think of it as a particle orbiting the atomic nucleus, Heisenberg imagined the electron as a probability wave.

hope to do is to say the electron is likely to be somewhere; we can never say with certainty that it is there. This is perhaps one of quantum mechanics' stranger ideas. How can a particle be perhaps here, but also perhaps there?

Rather than consider the electrons of an atom to be in fixed orbits as Bohr had done, Heisenberg conceived of them as representing the harmonics of a series of standing waves. He formulated equations that linked these waves to the quantum leaps of the electron from one orbit to another. Heisenberg shared his calculations with fellow physicist Max Born, who saw that Heisenberg's idea could be set out mathematically to link the energies of electrons to the emission lines that had been observed in the visible light spectrum. Born described the electron wave as being like a graph mapping out the probability of finding the electron in a particular place.

Niels Bohr declared that it was meaningless to ask what an electron really is. Experiments designed to measure waves will see waves, while experiments designed to measure the properties of particles will see particles. It is impossible to design an experiment that would allow us to see wave and particle at the same time.

SUPERPOSITION AND QUANTUM CATS

According to quantum theory, until it is measured a quantum system simultaneously exists in all possible states and positions. This is called superposition. When you spin a coin and then cover it with your hand you don't know how the coin has landed but you know that it is either 'heads' or 'tails'. In a quantum superposition state, the coin would be both at the same time. The system's superposition state is described by its wavefunction; the act of measurement, or observation, collapses the wavefunction, causing the system to take up a definite value for the property being measured.

Erwin Schrödinger came up with a thought experiment that would pass into quantum legend. 'A cat is penned up in a box,' Schrödinger wrote, 'along with the following device: in a Geiger counter there is a tiny bit of radioactive substance, so small, that perhaps in the course of the hour one of the atoms decays, but also, with equal probability, perhaps none: if it happens... a relay releases a hammer which shatters a small flask of hydrocyanic acid.'

He explained that the wavefunction of the entire system would express the situation by having in it the living or dead cat 'mixed or smeared out'. Schrödinger and Albert Einstein believed that this thought experiment had demonstrated that there was something distinctly not right about the Copenhagen Interpretation. Einstein said that a wavefunction that 'contains the living as well as the dead cat just cannot be taken as a description of a real state of affairs.'

The Copenhagen Interpretation

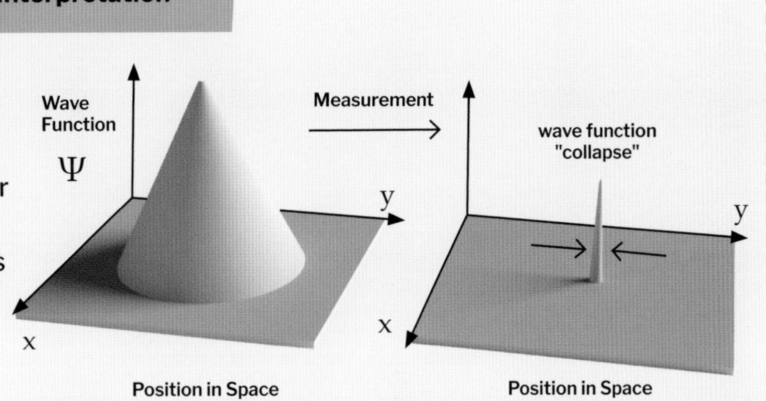

The Copenhagen Interpretation of quantum physics postulates that a quantum system exists in all possible states until it is observed. Observation collapses the system's wavefunction, causing it to take up a definite state.

Niels Bohr and Werner Heisenberg brought together their ideas on quantum physics in what became known as the Copenhagen Interpretation.

The quantum world as seen through the Copenhagen Interpretation is one of pure statistical probability. The Copenhagen view is that indeterminacy is a fundamental feature of nature and not just something that results from our lack of knowledge. We have to accept that this is how things are and not try to explain them. It opened up a sharp divide between the deterministic world of classical physics, where every event was assumed to have a cause, and the new quantum world of chance and uncertainty.

One of the central strands of the Copenhagen Interpretation is the principle of complementarity. This views the wave and particle nature of objects as complementary aspects of a single reality. Just as a tossed coin can come down heads or tails, an electron or a photon, for example, can sometimes be a wave and sometimes a particle, but never both at the same time. The Copenhagen Interpretation treats Schrödinger's wavefunction as no more than a tool for predicting the results of observations, and cautions physicists against concerning themselves with trying to imagine what 'reality' looks like.

The chance nature of the Copenhagen Interpretation was not without its opponents, including Albert Einstein who held fast to the notion that there was an objective reality that could, in theory, be measured. In an often-quoted letter to Max Born, written in 1926, Einstein wrote: 'Quantum mechanics is certainly imposing. But an inner voice tells me that it is not yet the real thing… I, at any rate, am convinced that He does not throw dice.'

Erwin Schrödinger's famous 'alive-or-dead' cat was a challenge to the Copenhagen Interpretation.

Niels Bohr saw no reason why the rules of classical physics, which determine what goes on in the everyday world around us, should also apply to the quantum realm. What the quantum physicists were discovering was just the way things were, whether Einstein and Schrödinger liked it or not. At some point, an exasperated Bohr apparently declared: 'Stop telling God what to do!'

28

Spooky action

Quantum entanglement and the EPR paradox

It is a basic tenet of quantum mechanics that we cannot measure all of the features of a system simultaneously – an idea enshrined in Werner Heisenberg's uncertainty principle. The Copenhagen Interpretation says that the very act of measurement selects the characteristics observed. Furthermore, the measurement of one particle having an instantaneous effect on another particle far distant from it.

THE PHENOMENON OF QUANTUM ENTANGLEMENT

If two electrons, for example, are ejected from a quantum system, then conservation of momentum laws tell us that the momentum of one particle is equal and opposite to that of the other. The two electrons are linked by a single wavefunction. The electrons are said to be 'entangled', and even though they are separated they will continue to share this single wavefunction. According to

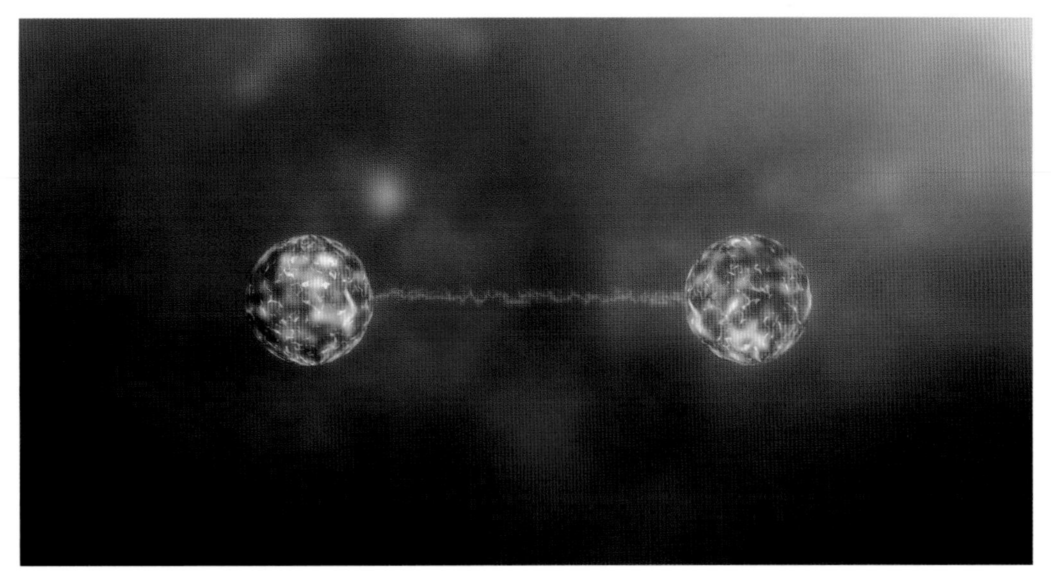

Quantum entanglement means that two particles remain connected in some way, even over great distances.

the Copenhagen Interpretation, until we measure it we have no idea what the momentum of either particle is, but because they are entangled, measuring the momentum of one collapses the wavefunction and determines the state of the other, fixing its momentum too, even if the electrons are separated by light years of space. This is known as 'non-local behaviour', but Albert Einstein called it 'spooky action at a distance'.

Einstein objected on the grounds that entanglement would require a signal to pass from one entangled particle to the other at a speed faster than that of light, something that was expressly forbidden by his theory of special relativity. Einstein accepted that quantum mechanics wasn't 'wrong' as such – its ability to predict experimental results was undeniable – but he was convinced that it was incomplete and there were still properties to be discovered, which he referred to as 'hidden variables'. Together with co-authors Boris Podolsky and Nathan Rosen, Einstein set out to demonstrate this by means of what became known as the EPR paradox.

EINSTEIN VERSUS BOHR

If there was any way that we could learn with absolute certainty the position of a particle, Einstein and co maintained, and we don't disturb the particle by directly observing it, then we can say that the particle exists in reality, independent of our observations. If we can take measurements of this particle that give us information about a second particle without disturbing the second particle in any way – for example, measuring the momentum of the first particle gives us precise knowledge of the momentum of the second particle – it means that the second particle, which we have not directly observed, has properties that we know. It has a momentum that is real. Einstein and his collaborators argued that the assumption was being made that the process of measuring the first particle instantaneously alters the reality of the second particle, even if they were separated by light years of space, something they believed that 'no reasonable definition of reality could be expected to permit'.

Bohr disagreed and passionately defended the Copenhagen Interpretation of quantum mechanics. He rejected Einstein's thought experiments by recourse to the uncertainty principle. If the two particles are entangled, Bohr argued, then they are effectively a single system that has a single quantum function. It is not possible to know both the precise position and the precise momentum of either particle at the same moment. If you know the position of A then you know the position of B, and if you know the momentum of A you know the momentum of B. But it is still impossible to know both position and momentum for either particle. There is no conflict with the uncertainty principle.

Einstein continued to counter that quantum mechanics violated two

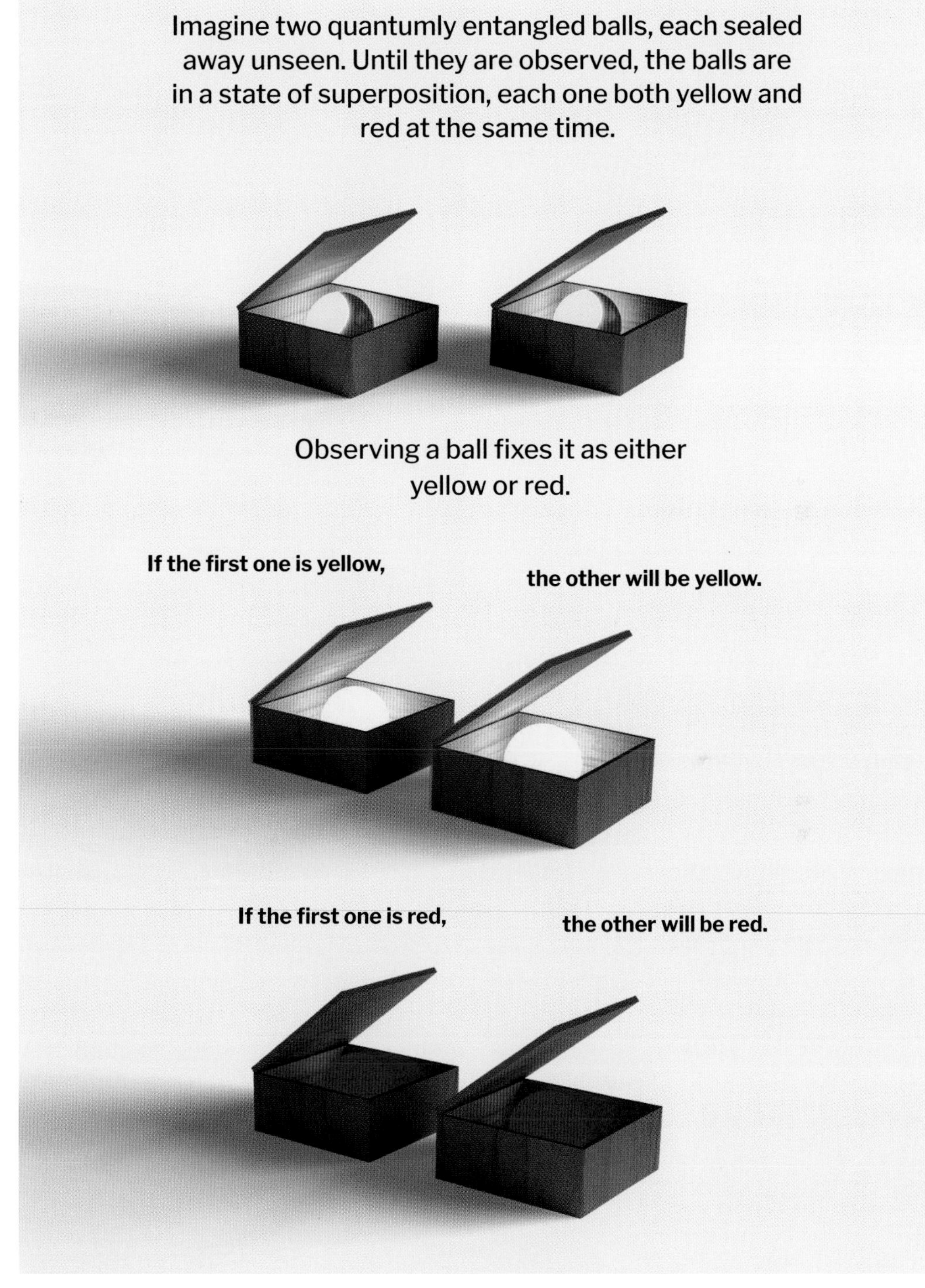

Observing one quantumly entangled particle instantly affects the state of the other particle, no matter how far away it is.

fundamental principles: the principle of separability, which maintains that two systems separated in space have an independent existence; and the principle of locality, which says that doing something to one system cannot immediately affect the second system.

John Stewart Bell demonstrated mathematically that quantum entanglement is a real phenomenon.

BELL'S THEOREM

In 1964, physicist John Bell proposed an experiment that could test whether or not entangled particles actually did communicate with each other faster than light. According to quantum theory, entangled particles remain in a superposition of states until they are measured, but as soon as one is measured, we know with certainty that the other has to have the complementary characteristic. Bell derived formulas, called the Bell inequalities, which determine how often the characteristics of particle A should correlate with those of particle B if normal probability (as opposed to quantum entanglement) was at work in determining their states. Bell proved mathematically that the predictions of quantum theory were indeed not in line with those of normal probability and that there is indeed an instantaneous connection between entangled particles. In the words of physicist Fritjof Capra, 'Bell's Theorem demonstrates that the universe is fundamentally interconnected.'

Experiments such as those carried out by French physicist Alain Aspect in the early 1980s, which used entangled photon pairs generated by laser, have demonstrated convincingly that 'action at a distance' is real. Aspect found that the measurements made of the entangled pairs correlated 40 times more often than would have been expected if normal probability had applied. The quantum realm is not bound by the rules of locality. When two particles are entangled, they are effectively a single system that has a single quantum function.

29

QED

Quantum field theory

A field can be thought of as anything that has values that vary across space and time. Electromagnetism and other fundamental forces arise from variations in the fields that carry them. In the 1920s, quantum field theory proposed a different approach, suggesting that forces were carried by means of quantum particles, such as photons.

One of the biggest shortcomings of quantum mechanics is that it fails to take Einstein's relativity theories into account. One of the first to try reconciling those cornerstones of modern physics was Paul Dirac. The Dirac equation, which viewed electrons as excitations of an electron field, in the same way that photons could be seen as excitations of the electromagnetic field, became one of the foundations of quantum field theory.

QUANTUM ELECTRODYNAMICS

Quantum electrodynamics, usually referred to as QED, is the quantum field theory that deals with the electromagnetic force. The QED theory was fully developed in the late 1940s by Americans Richard Feynman and Julian Schwinger, and Japanese physicist Shin'ichirō Tomonaga, who were all working independently of each other.

QED proposes that charged particles such as electrons interact with each other by emitting and absorbing photons, the force carriers of the electromagnetic force. These photons are 'virtual' particles, which means that they cannot be seen or detected in any way. They simply represent the force of the interaction between the charged particles, which causes them to change their speed and direction of travel as a consequence of releasing or absorbing the photon's energy. According to QED theory, the more complex the interaction the greater the number of virtual photons that are exchanged in the process and the less likely it is to occur.

The ways in which particles can interact by the exchange of photons can be visualized by means of Feynman diagrams, developed by Richard Feynman. Each diagram represents the particles involved and their interactions. Feynman diagrams have proved to be invaluable in helping scientists tackle some of the complex interactions

Quantum field theory has been described as 'physics in a bad mood'. It suggests that forces are transmitted by quantum particles.

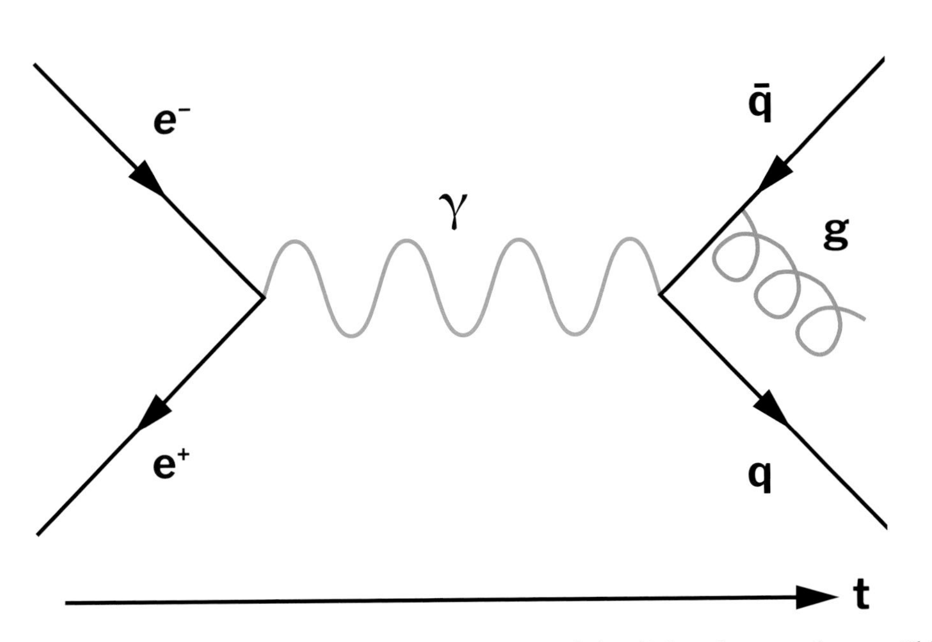

Richard Feynman developed Feynman diagrams as a way of visualizing photon exchanges. This example shows an electron/positron pair annihilating to produce a virtual photon that becomes a quark/antiquark pair. The antiquark then radiates a gluon.

involved in high energy physics.

QED is one of the most astonishingly accurate theories ever produced. QED's prediction for the strength of the magnetic field associated with an electron is so close to the value determined by experiment that if the distance from London to Timbuktu was measured to the same precision it would be accurate to within a hair's breadth.

Stepping stone

Quantum electrodynamics was the first theory that attempted to combine quantum mechanics with Einstein's special relativity. The success of QED was a stepping-stone towards building quantum field theories for the other fundamental forces of nature. Thanks to Einstein and the theory of general relativity (see page 167), physicists have a workable explanation for gravitational forces as resulting from a curvature of spacetime brought about by the matter in it. In theory at least, an explanation involving an exchange of force particles, called gravitons, is equally valid. The problem is that there is no current quantum theory of gravity involving gravitons that is as well worked out and proven by experiment as Einstein's relativity theory and, as yet, no experimental proof of their existence.

30

Matter transformed

Discovering radioactivity

Before the end of the 19th century scientists believed that matter would only emit radiation, such as visible or ultraviolet light, if it was stimulated in some way, such as by heating. This supposition was disproved by investigations into x-rays.

In 1895, German physicist Wilhelm Röntgen was experimenting with cathode ray tubes in his laboratory, when he noticed that a nearby fluorescent screen had started to glow. He concluded that a new type of ray was being emitted from the tube. Later experiments established that this mysterious 'x-ray' could pass

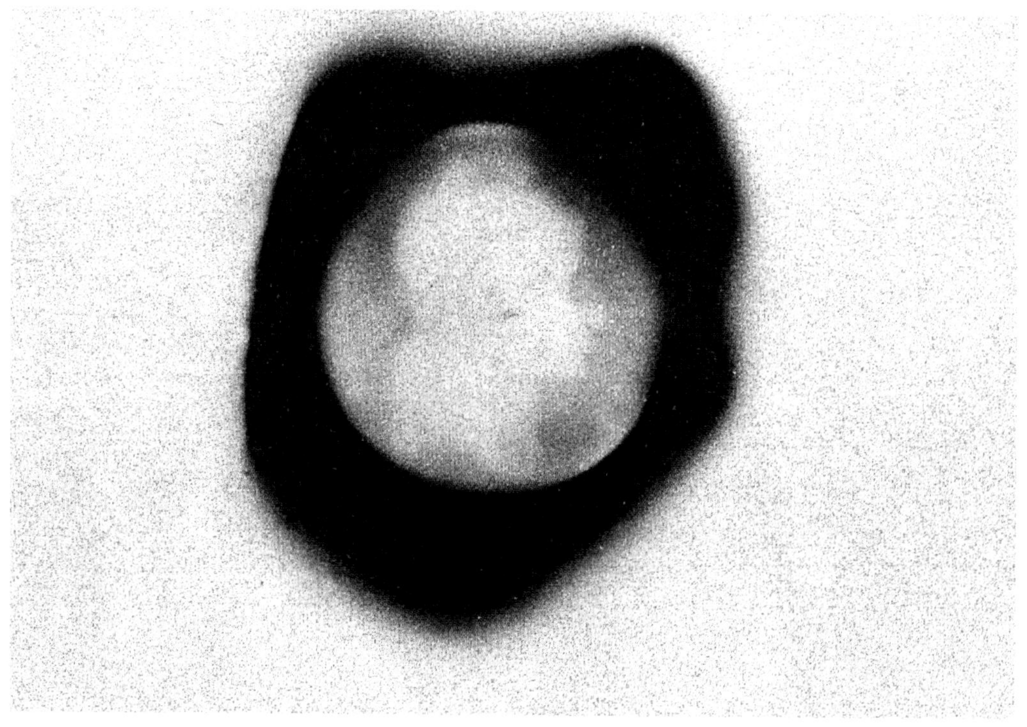

Henri Becquerel's photographic plates, darkened by exposure to uranium, were evidence of the existence of radioactivity.

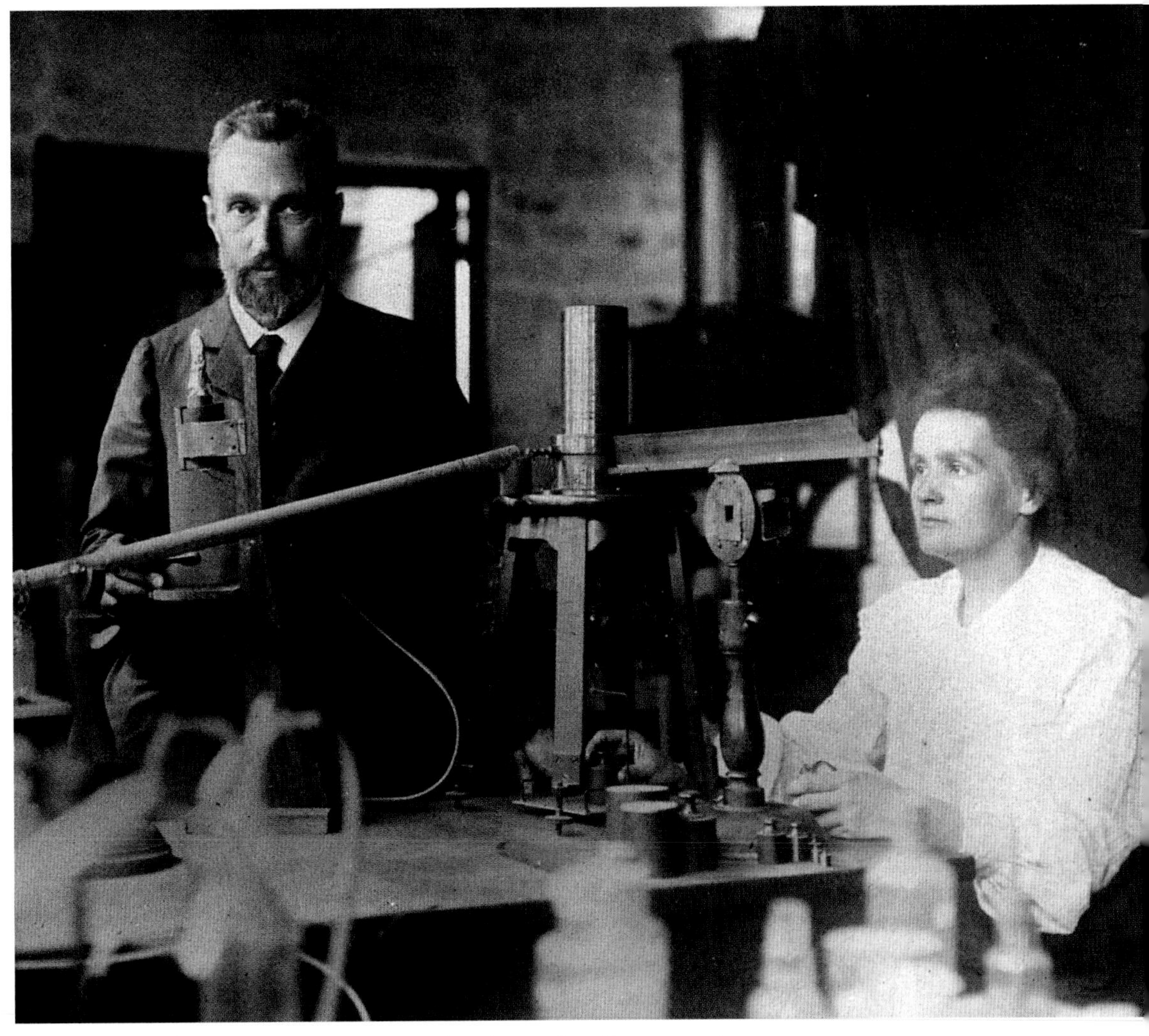

Pierre and Marie Curie spent years investigating radioactive elements.

through most substances, including through the soft tissue of humans, but not bones and metal objects. The discovery, which would win Röntgen the first Nobel Prize for Physics in 1901, caught the attention of both scientists and the general public around the world, who were fascinated by the invisible ray.

One such investigator was French physicist Henri Becquerel. He believed that uranium absorbed the sun's energy and then emitted it as x-rays. He exposed a uranium compound to sunlight and then placed it on photographic plates wrapped in black paper. His experiment was frustrated because the day was overcast, but Becquerel decided to develop his photographic plates anyway. To his surprise, clear outlines of the compound were formed on the plate.

Becquerel concluded that the uranium must be emitting radiation on its own without external stimulation from the sun. He found that the energy emitted by the uranium compound appeared not to diminish, even over a period of several months.

DISCOVERIES OF THE CURIES

Further investigations by Becquerel and others, such as Marie Curie and her husband Pierre, revealed that other substances had similar properties. Marie Curie coined the word 'radioactivity' to describe Becquerel's discovery. The Curies discovered that samples of pitchblende, a mineral that contains uranium, appeared to produce even more radioactivity than pure uranium and concluded that there had to be another radioactive substance present. Eventually they isolated a sample of a new chemical element, over 300 times more radioactive than uranium, which they called polonium. They were surprised to find that the waste left behind after the polonium had been extracted remained highly radioactive. Further years of arduous work led to the discovery of another new element, radium. In 1903, Marie and Pierre Curie were jointly awarded the Nobel Prize for Physics with Henri Becquerel for their work on radioactivity.

Marie Curie observed that an ounce of radioactive radium would produce 4,000 calories of heat per hour, seemingly indefinitely. The source of this energy was explained a few years later in 1905 by Albert Einstein and his theory of special relativity. According to Einstein, mass and energy are equivalent, as summed up in the iconic $E=mc^2$ equation. As the radium radiated heat it was also losing a tiny amount of mass, which was being converted into energy.

RUTHERFORD'S RADIATION

In 1898, New Zealand-born Ernest Rutherford discovered that there were at least two distinct types of radiation which, simply for convenience, he designated α (alpha) and β (beta). He established that the beta rays were a hundred times more penetrating than the alpha rays. Further investigation established that beta rays were deflected by a magnetic field, indicating that they were negatively charged electrons. In 1903, Rutherford discovered that alpha rays were deflected slightly in the opposite direction, suggesting that they were massive, positively charged particles. Rutherford later proved, in 1908, that alpha rays are in fact the nuclei of helium atoms.

In 1900, a third type of radiation was identified by French chemist Paul Villard. Dubbed γ (gamma) rays, these were even more penetrating than alpha rays. Gamma rays were later shown to be a high-energy form of electromagnetic radiation, like x-rays but with a much shorter wavelength.

Rutherford and chemist Frederick Soddy discovered in 1901 that one

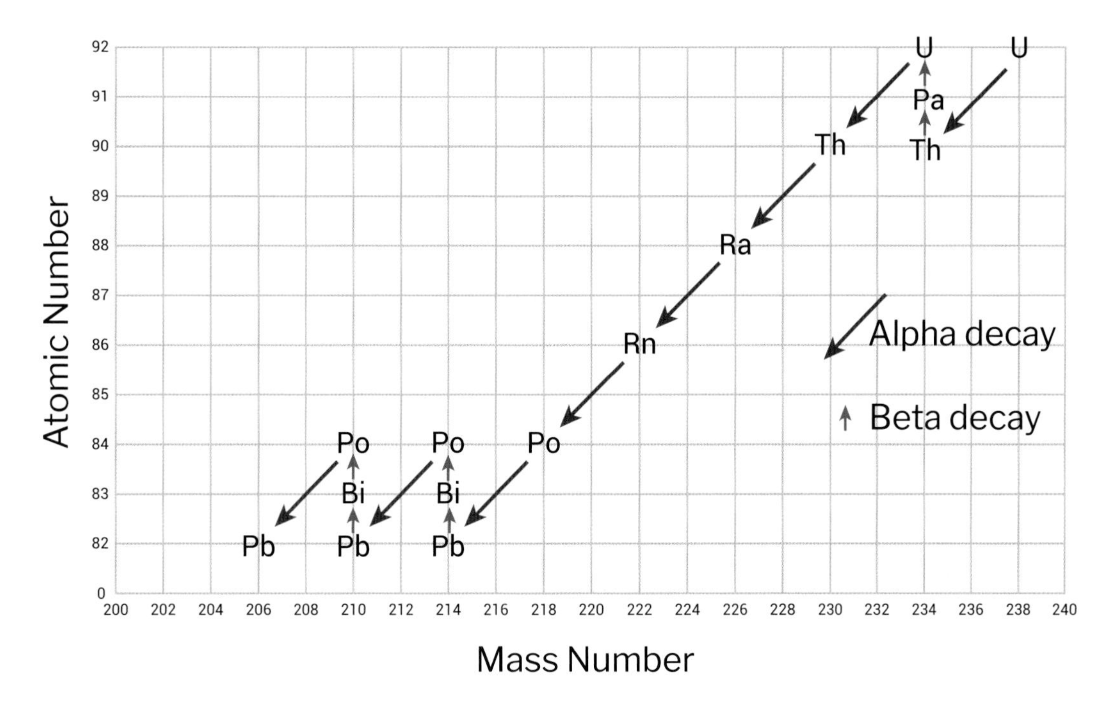

The radioactive isotope uranium 238 eventually decays into stable non-radioactive lead through a long chain of transformations into different elements.

radioactive element transforms, or decays, into another. The elements of the periodic table can come in more than one form, called isotopes, some of which are more stable than others. Not surprisingly, the most stable form of an element is generally the most common in nature. All elements have an unstable form which is radioactive and emits radiation. Some elements, such as uranium, have no stable form and are always radioactive. These unstable elements decay, transforming into different elements, called decay products. While doing this they emit radiation in the form of alpha and beta particles and gamma rays. If the decay product is itself unstable the process continues until a stable, non-radioactive, form is reached.

The individual atoms in the radioactive substance change randomly, but at a characteristic rate that depends on the element involved. This came to be termed the half-life of a radioactive element – the time it takes for one half of the radioactive sample to decay. The half-life of different elements can range from billions of years to a tiny fraction of a second.

31

Atomic building blocks

The structure of the atom

The idea that matter is made up of elementary particles stretches back to ancient Greece and the indivisible, indestructible atoms of Democritus and was revived in the atomic theories of John Dalton and Amedeo Avogadro at the beginning of the 19th century. Since its existence was first proposed, the atom has gradually revealed its often surprising and unexpected nature.

DISCOVERING THE ELECTRON

In 1858, German physicist Julius Plücker sent a high voltage between metal plates inside a glass tube from which most of the air had been removed and observed a green glow inside the tube near the cathode. English physicist William Crookes carried out further investigations of these 'cathode rays' in 1879 and found that they could be bent by a magnetic field and were apparently made up of negatively charged particles.

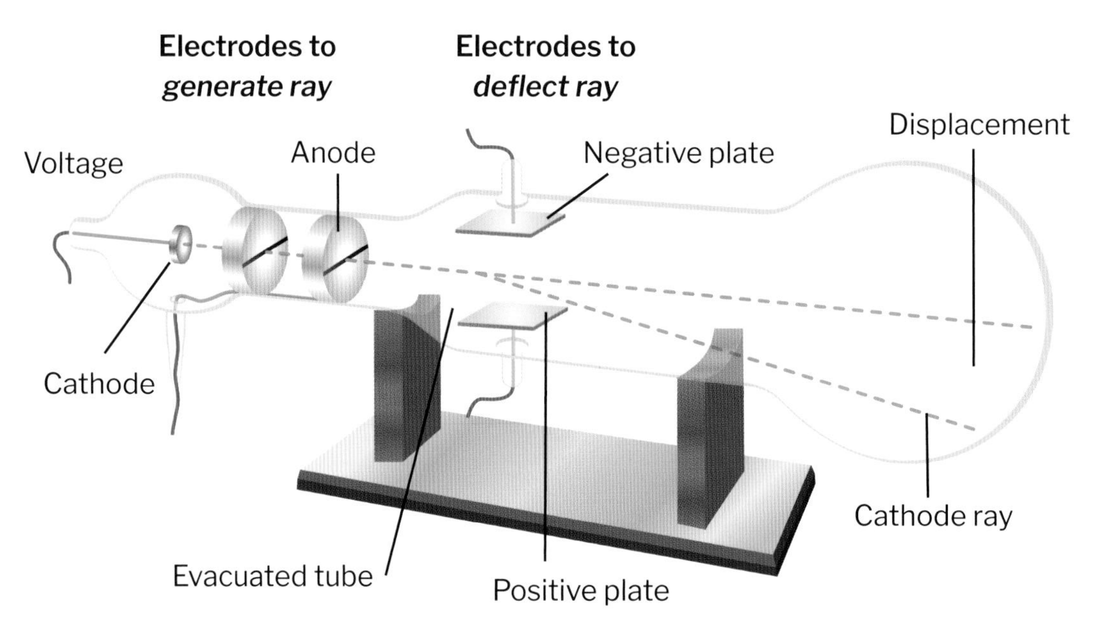

Diagram of the apparatus used by Thomson to detect the electron.

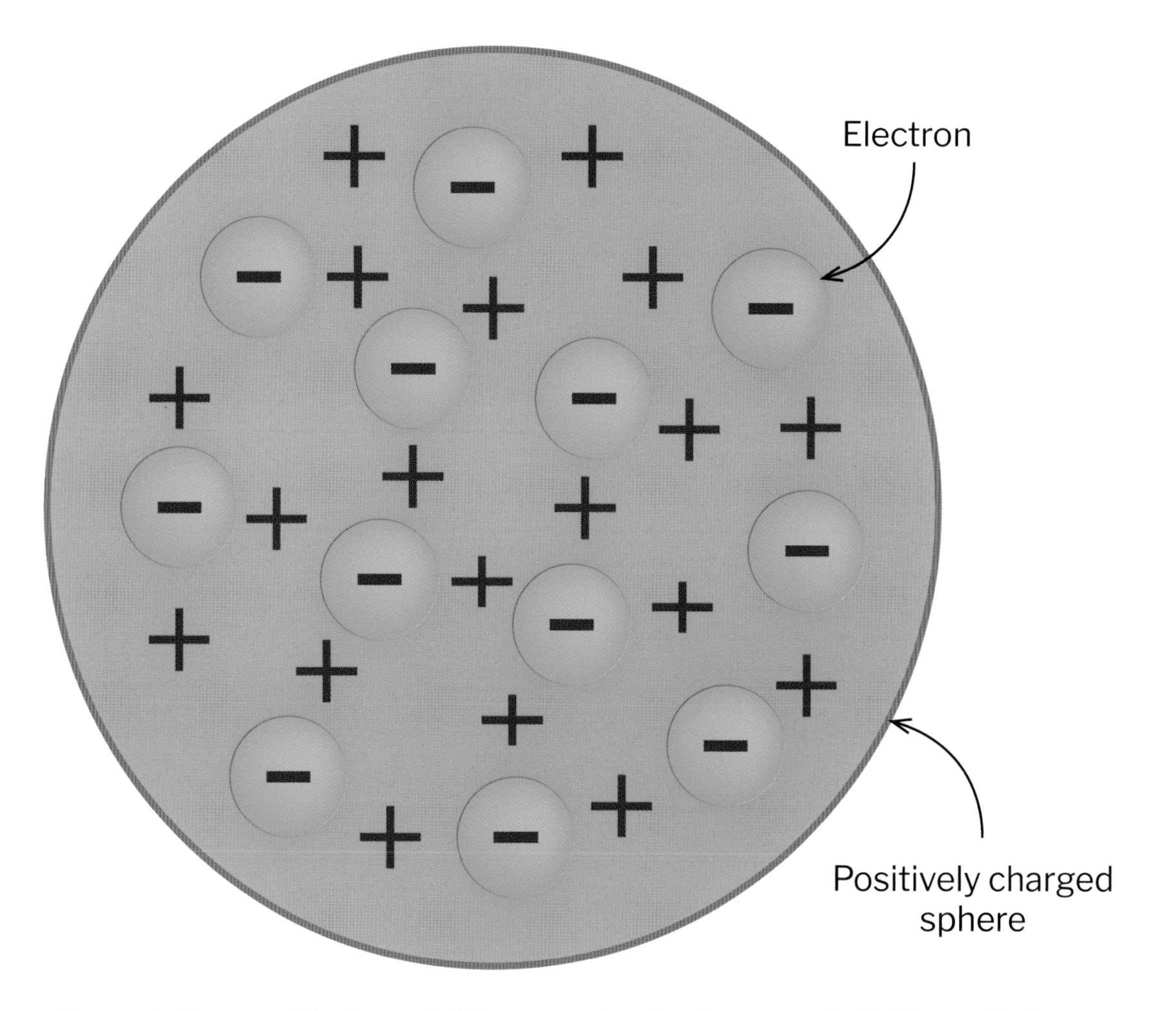

Thomson's discovery of the electron led him to speculate that they were embedded in a positively charged cloud surrounding the atom, like raisins in a plum pudding.

In 1894, British physicist J.J. Thomson began a series of experiments that would settle the nature of the cathode rays once and for all. He discovered that cathode rays could also be deflected by an electric field and determined the ratio of the charge of the mystery particle to its mass. Thomson found the charge-to-mass ratio remained the same regardless of the metal used to make the electrodes, or of any gas in the tube. He deduced that the particles making up the cathode ray must be something that was found in all forms of matter. By 1897, Thomson had determined that the negatively charged particles of the cathode ray had a mass that was less than 1/1,000th that of a hydrogen atom. They could not be any particle then known to physics. The name 'electron', which had been proposed by Irish physicist George Stoney for the fundamental unit of electrical charge, was soon adopted.

How did the electron fit in to the structure of the atom? It was known that atoms were electrically neutral, so Thomson proposed that the negatively charged electrons were embedded in a positively charged cloud, like raisins in a cake, an image that led to Thomson's atom being dubbed the 'plum pudding' model. Within a few years, however, new discoveries showed up the flaws in the pudding.

THE NUCLEUS

In 1909, Ernest Rutherford set student Ernest Marsden the task of carrying out an experiment that involved firing alpha particles from a radioactive source at a thin gold foil. According to Thomson's plum pudding model, the positive charge of the 'pudding' should be widely distributed throughout the volume of the atom. The large, fast-moving alpha particles should have passed through the positive pudding in the gold foil with scarcely any deflection as the electric field in the atom was too weak to affect them. But instead, Marsden saw evidence of the alpha particles being repelled. As an astonished Rutherford later remarked, 'It was as if you fired a 15-inch shell at a piece of tissue paper and it came back and hit you.'

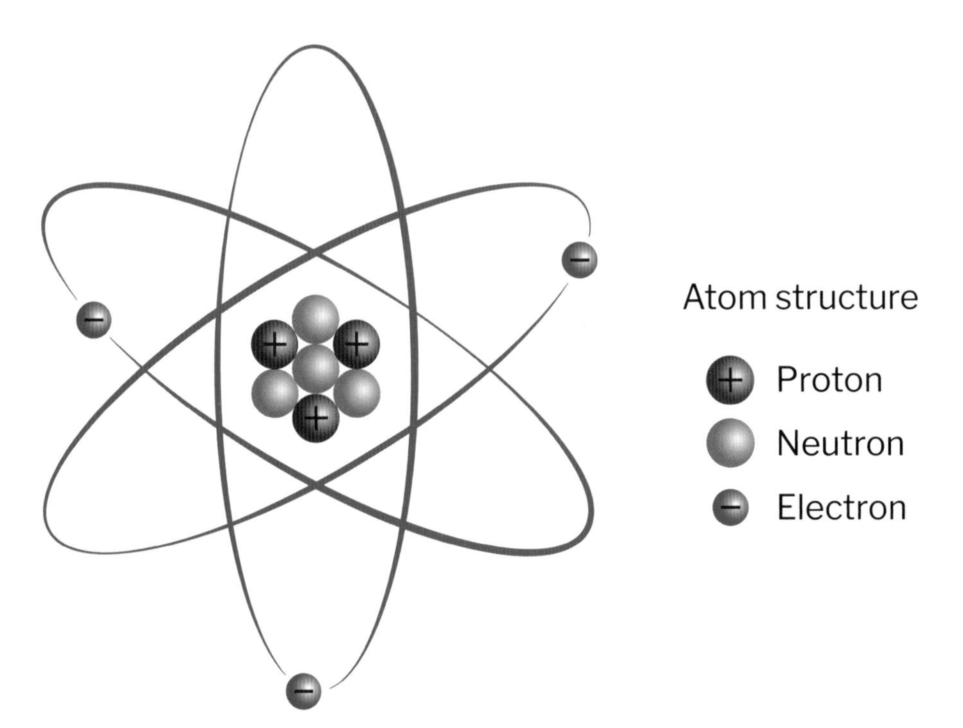

The planetary model of the atom.

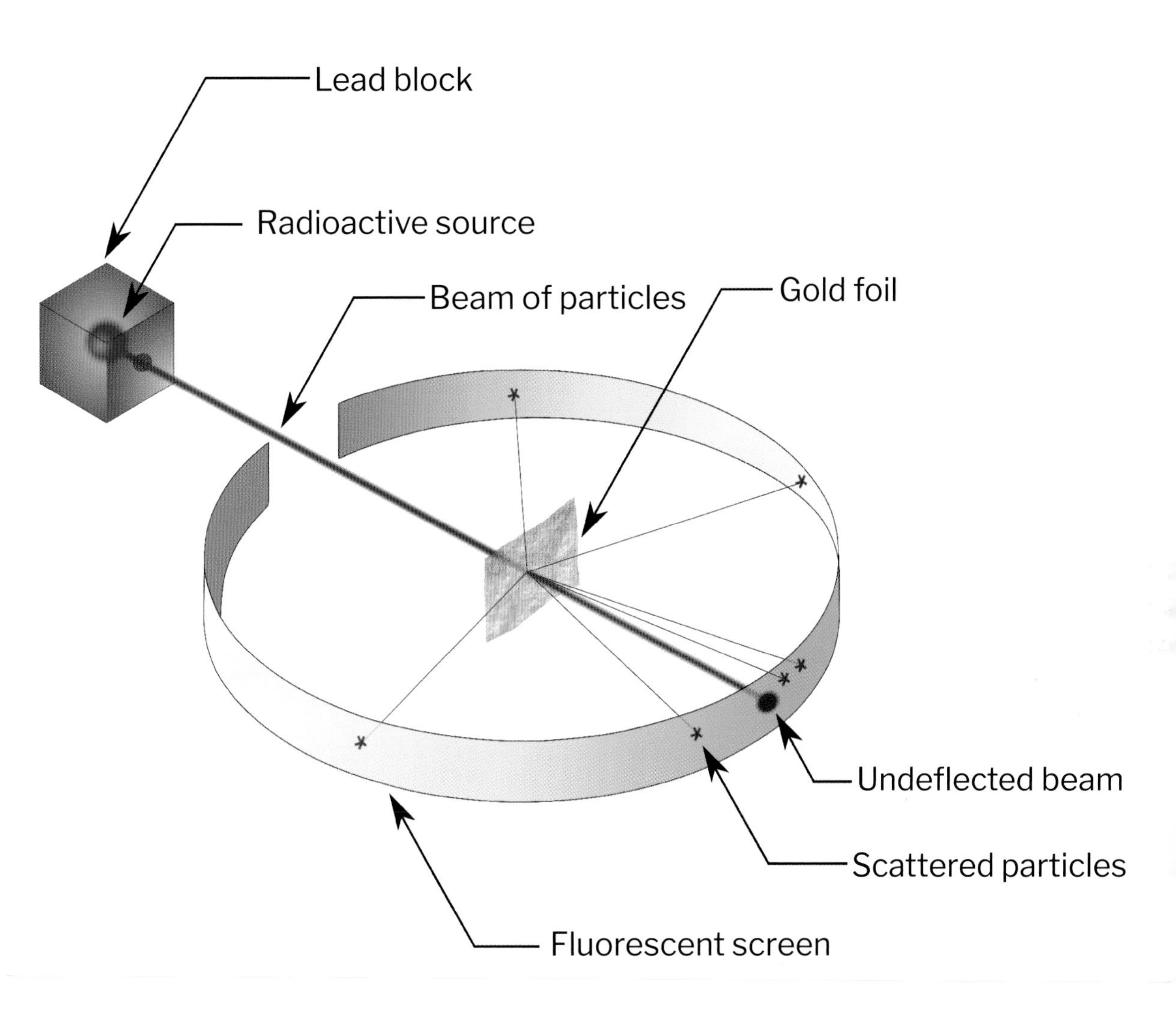

Ernest Rutherford and Ernest Marsden were astonished to discover that there was a positively charged nucleus at the centre of the atom.

Rutherford concluded that the alpha particles must have been repelled by large positively charged particles and proposed a new model of the atom in 1911. He suggested that most of the mass of the positive charge is concentrated in a nucleus situated in the middle of the atom, with the electrons orbiting this central core, like planets around a star. Rutherford calculated the size of the nucleus and found it to be only about 1/100,000th the size of the atom. The atom, it appeared, was mostly empty space.

At the heart of the atom was a particle with a positive charge, which Rutherford named the proton, from the Greek

James Chadwick.

word 'protos', meaning 'first'. Atoms of different elements have different numbers of protons, with the hydrogen nucleus, the smallest atom, having a single proton. The atomic number of an atom equals the number of protons in the nucleus, equivalent to the positive charge of the atom. However, research showed that the atomic number was less than the atom's atomic mass. A helium atom, for instance, has an atomic mass of 4, but an atomic number (or positive charge) of 2. There had to be something besides the protons in the nucleus to account for the missing mass. Rutherford suggested that there could be a particle, consisting of a paired proton and electron, which he called a neutron, that had a mass similar to that of the proton but no charge. In 1932, James Chadwick, a former student of Rutherford's, discovered a neutral particle similar in mass to the proton. Werner Heisenberg showed that the neutron could not be a proton-electron pairing but was a unique fundamental particle in its own right. The discovery changed physicists' conception of the atom and they soon found that the chargeless, massive neutron made an ideal projectile for bombarding the nucleus. Before long, neutron bombardment of the uranium atom was being used to split its nucleus and release the huge amounts of energy predicted by Einstein's $E=mc^2$, opening up the possibility of atomic energy and the atomic bomb.

32

Muster Mark's Quarks

The Standard Model and the particle zoo

By 1932, physicists had established that atoms were formed from three particles – electrons, protons and neutrons – but questions remained. What held the nucleus together? Why didn't the positively charged protons in the nucleus push each other apart? The discovery of more subatomic particles further complicated things. How could it all be brought together in a coherent whole?

In August 1912, Austrian physicist Victor Hess ascended to 5,300 metres (17,400 ft) in a balloon to take measurements of radiation in the high atmosphere. He discovered that at that altitude it was some three times what it was at sea level. He had discovered cosmic rays – high-energy particles entering the atmosphere from outer space.

The discovery opened up a previously unknown world of subatomic particles. In 1932, Carl Anderson was studying cosmic particles in a cloud chamber at the California Institute of Technology when he spotted something with the same mass as an electron, but positively charged (see page 145). Further observations led him to conclude the tracks were actually due to antielectrons, produced alongside an electron from the impact of cosmic rays in the cloud chamber. He called the antielectron a 'positron'.

In 1936, while studying cosmic radiation cloud chamber trails, Anderson noticed signs of negatively charged particles that curved more sharply than electrons, but less than protons when passing through a magnetic field. He concluded that the new particle must be somewhere between the electron and the proton in mass. The mesotron, as Anderson called it, was later renamed the mu meson, adopting the more general term meson to refer to any particle with a mass intermediate between that of electrons and that of protons and neutrons.

The existence of the neutrino was first predicted in 1931 when Wolfgang Pauli theorized the existence of an unknown particle to account for an apparent loss of energy and momentum he detected in his studies of radioactive decay. The elusive particle was first detected in 1959 in the course of experiments at a nuclear reactor in South Carolina. We now know that

The Cloud Chamber

The invention of the cloud chamber gave researchers a way to track the movements of subatomic particles. The first cloud chamber was built by physicist Charles Wilson in 1911. Air inside the chamber was saturated with water vapour and then the pressure was lowered. When, for example, a positively charged alpha particle moved through the chamber it removed electrons from the gas, leaving charged atoms which attracted the water vapour and caused a visible trail to form. Used together with magnets and electric fields, the cloud chamber gave physicists the ability to calculate properties such as mass and charge.

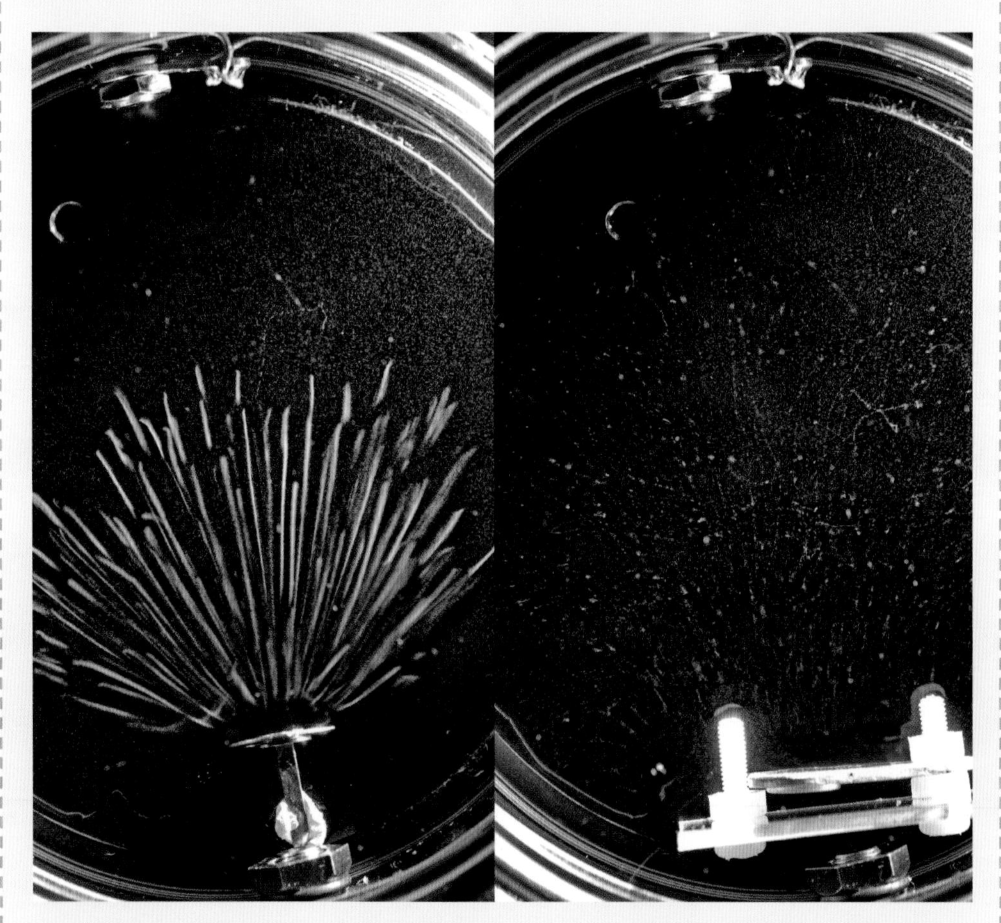

Tracks produced by alpha radiation (left) beta radiation (right) in a cloud chamber. The cloud chamber was an invaluable tool for investigating the properties of atomic particles.

Particle accelerators

Particle accelerators are one of the foremost tools for exploring the subatomic world. They are all based on a fairly simple principle. In the same way that a falling object accelerates under the force of gravity, so a charged particle accelerates across a potential difference (voltage). In a sense, the particle 'falls' through the potential difference.

The first cyclotron, described as a 'proton merry-go-round', was built in 1931. This was a circular accelerator in which the electric field was used to accelerate the particles more than once. A magnetic field sent the accelerated particles along a spiral path, taking them back and forth between two electrodes, so they were accelerated again and again until they were eventually emitted from the cyclotron as a high-energy beam.

The advantages of linear accelerators are that they are capable of accelerating larger ions than circular accelerators and it is easier to produce high-energy electron beams. They are widely used today in research and in medicine. The 3 km long (1.8 miles) Stanford Linear Accelerator Centre is capable of accelerating electrons and positrons to energies of 50 GeV (50,000,000,000 eV).

The most powerful of the particle accelerators in operation today is CERN's Large Hadron Collider. The LHC is a synchrotron, accelerating particles around a circle 27 km (17 miles) long at speeds of up to 99.999999 per cent of the speed of light.

The Large Hadron Collider at CERN can accelerate particles to close to the speed of light.

neutrinos are everywhere, permeating everything in the universe. Every second over 100 billion neutrinos pass through every square inch of your body. They barely interact with any other particles and can pass unperturbed right through the earth.

By the 1960s hundreds of new subatomic particles had been discovered. Physicists began sorting this burgeoning particle zoo into more manageable categories. There were the hadrons, which included the baryons, heavier particles such as the proton and neutron and their corresponding antiparticles, as well as the intermediate mesons; and the lighter leptons, such as the electron and the neutrino.

In 1962, Murray Gell-Mann and Yuval Ne'eman proposed a possible solution to the perplexing proliferation of particles. They devised a scheme for classifying the hadrons called the Eightfold Way, based on a mathematical symmetry known as SU(3) and named after Buddha's Eightfold Path to enlightenment.

The Eightfold Way, rather like chemistry's periodic table of the elements, could be used to describe and categorize the characteristics of known particles, such as the magnetic properties of protons, and predict the properties of particles that hadn't even been seen yet. Underpinning all of this were three new elementary particles, which Gell-Mann named quarks, taking the name from a passage in James Joyce's novel *Finnegan's Wake* – 'Three quarks for Muster Mark!' Experiments at the Stanford Linear Accelerator in the late 1960s confirmed the existence of quarks.

Quarks, along with leptons, appear to be the true building blocks of matter. There are believed to be six 'flavours' of quark – up, down, charm, strange, top and bottom – combinations which can successfully account for the more than 200 types of meson and baryon known to physics. Quarks have fractional electric charge. The up quark, for example, has a charge of +2/3 and the down quark a charge of -1/3. The familiar protons and neutrons are constructed from three up and down quarks (up-up-down – total charge +1 – and up-down-down – total charge 0 – respectively). The force binding quarks together, called the colour force, is so powerful that quarks have never been detected in isolation. The theory that describes the interactions between quarks is called quantum chromodynamics.

THE STANDARD MODEL

The Standard Model, developed in the early 1970s, is a mathematical model that aims to bring together all that we know about particles and forces into a coherent whole. It postulates that everything in the universe is formed from a few basic building blocks, called fundamental particles, with the interactions between them governed by four fundamental forces: gravity, the electromagnetic force, and the weak and strong forces, which are effective only at the level of subatomic

Standard Model of Elementary Particles

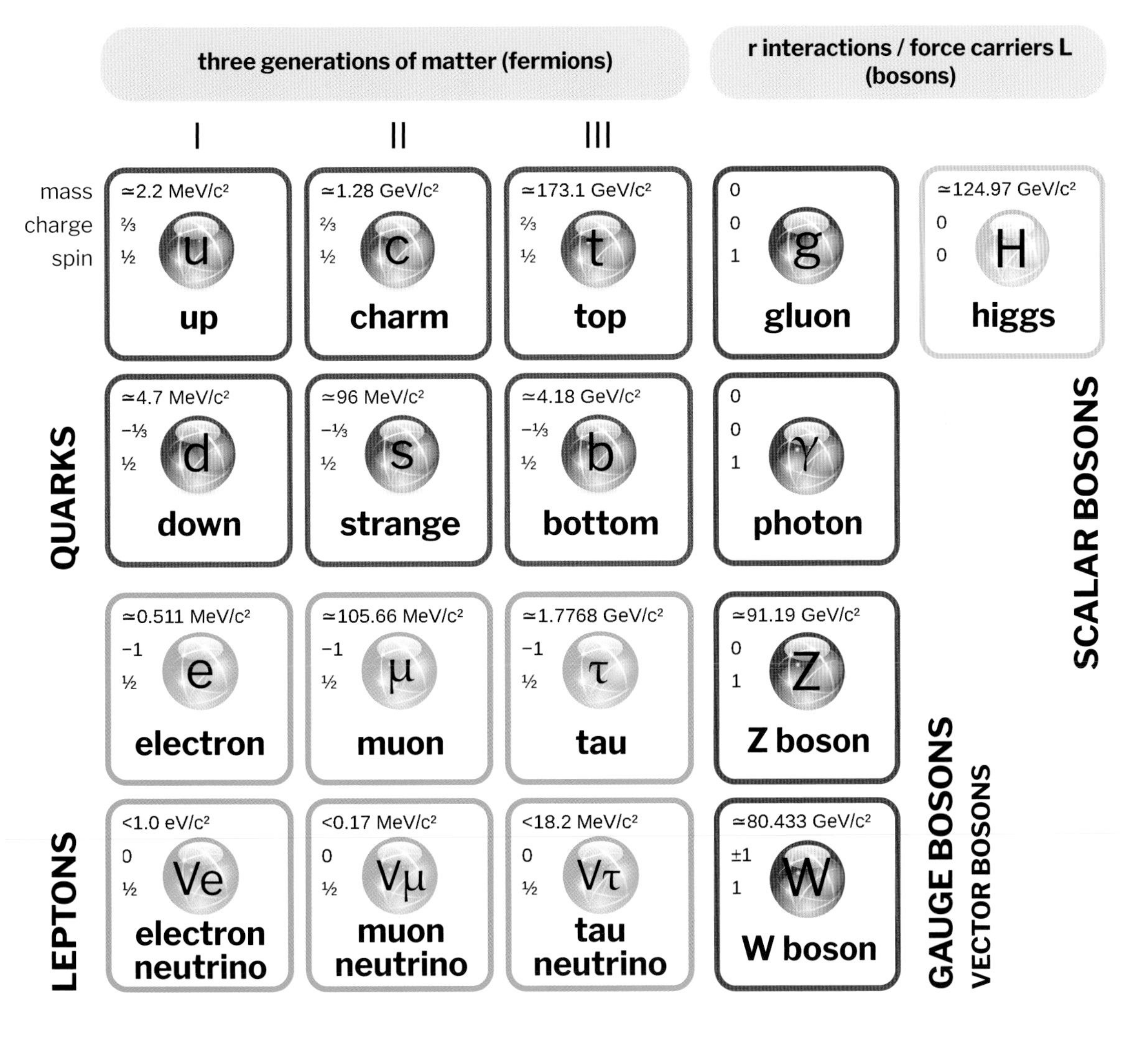

The Standard Model aims to group the fundamental particles that make up all matter in the universe along with the force-carrying particles that govern the interactions between them.

particles. Elementary particles are grouped into two classes: bosons, such as the photon, which transmit forces, and fermions, which include the quarks, the electron and the neutrino.

33

Force carriers

The fundamental forces

Four fundamental forces govern the behaviour of the matter in the universe – gravity, the electromagnetic force, and the weak and strong forces. The Standard Model is a tried and tested explanation for the way the fundamental particles and three of the four forces work together. The Standard Model explains forces as the result of matter particles exchanging bosons. The model is incomplete because it does not account for the force of gravity.

The four forces work over different ranges and have different strengths. Gravity is the weakest, but its range is infinite. The electromagnetic force also has infinite range but is many times stronger than gravity. The weak and strong forces are effective only at the level of subatomic particles; the weak force, despite its name, is much stronger than gravity but weaker than the electromagnetic force. The strong force, as the name suggests, is the strongest of the four.

Three of the fundamental forces result from the exchange of force-carrier particles, together called bosons. Each fundamental force has its corresponding boson – the strong force is carried by the gluon, the electromagnetic force is carried by the photon, and the 'W and Z bosons' are responsible for the weak force.

THE ELECTROMAGNETIC FORCE

The electromagnetic force has infinite range and acts between charged particles, such as negatively charged electrons and positively charged protons. Opposite charges attract, and like charges repel. The greater the charge, and the closer the charged particles, the greater the force it exerts. The electromagnetic force consists of the electric force and the magnetic force, at first thought to be separate but later realized to be twin aspects of the same force. Electromagnetic forces are transferred between charged particles through the exchange of photons, which are virtual and undetectable, even though they are technically the same particles as the real and detectable photons that are the particle form of light.

THE STRONG FORCE

In the Standard Model the basic exchange

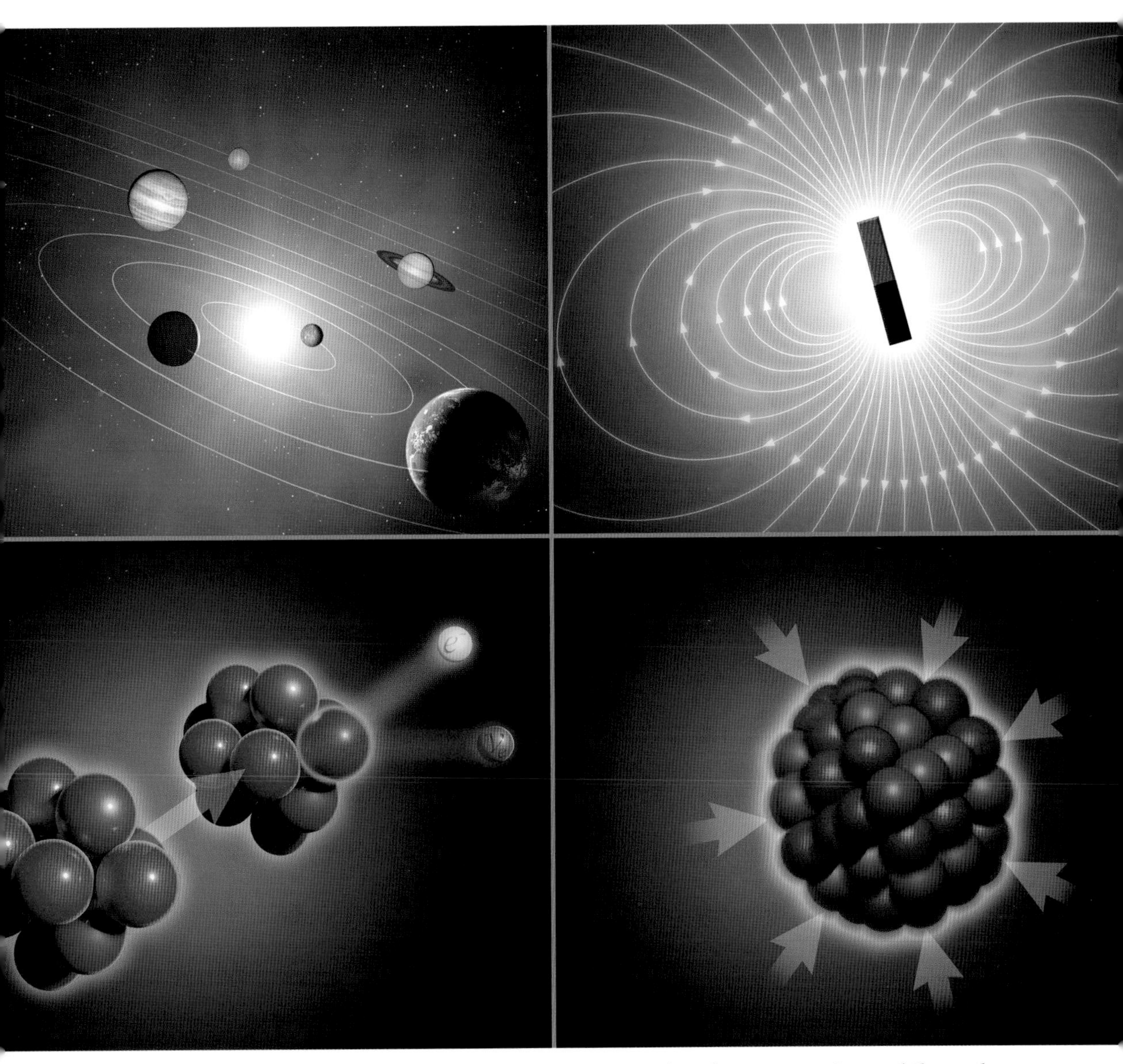

The four fundamental forces governing the universe are gravity, electromagnetism and the weak and strong nuclear forces.

particle is the gluon which mediates the colour force holding together the quarks that make up protons and neutrons. The strong nuclear force is believed to be a residual effect of the colour force that extends beyond the boundary of the proton or neutron. The strong force holds the nucleus of the atom together against the powerful force of repulsion trying to push the positively charged protons apart. Although the most powerful of the fundamental forces, it acts over a very short range, just over half the diameter of a proton, or 0.73 x 10-15 m.

THE WEAK FORCE

In 1933, Italian physicist Enrico Fermi devised a theory to explain beta decay, the process by which a neutron in a nucleus changes into a proton and expels an electron, called a beta particle in this context. He postulated a new type of force, the so-called weak interaction, that was responsible for the decay, transforming a neutron into a proton, an electron and a neutrino, though later research showed this to be an antineutrino. Fermi originally thought that the two particles actually had to be touching for the force to work, but the weak force has since been shown to be an attractive force that works at an extremely short range of about 0.1 per cent of the diameter of a proton. The weak force is carried by the W and Z bosons, which were discovered at CERN in 1983. W bosons are electrically charged – W+ (positively charged) and W− (negatively charged).

By emitting an electrically charged W boson, the weak force changes the flavour of a quark, i.e. changing an 'up' quark into a 'down' quark for instance, causing a proton to change into a neutron, or vice versa. The weak force plays a vital part in nuclear fusion, the process that powers the stars. The first step is to smash two protons together with enough energy to overcome the electromagnetic force pushing them apart, bringing them close enough for the strong force to bind them together. This creates an unstable form of helium, a nucleus with two protons, as opposed to the stable form of helium with two protons and two neutrons. Next, the weak force comes into effect and one of the pair of protons undergoes beta decay. A subsequent chain of reaction eventually leads to the formation of stable helium nuclei.

The Z boson is neutral and its interaction with particles is hard to detect. Experiments to find W and Z bosons led to a theory combining the electromagnetic force and the weak force into a unified 'electroweak' force in the 1960s. However, the theory required the force-carrying particles to be massless, and scientists knew that this couldn't be the case. Accounting for the missing mass would lead to the discovery of the Higgs boson (see page 147). Efforts to unite the electroweak force with the strong force in a single electronuclear force have so far failed.

GRAVITY

Gravity, the force we are perhaps most familiar with in everyday terms, does not fit easily into the Standard Model. Albert Einstein suggested in his theory of general relativity that gravity is not a force at all, but a consequence of the bending of spacetime by the matter embedded in it. Gravity may theoretically result from the exchange of massless gravitons but these have yet to be detected experimentally.

34

Broken atoms

Nuclear energy and nuclear bombs

The scientists exploring the inner workings of the atom at the beginning of the 20th century could scarcely have imagined that their discoveries would lead to a new source of energy and the terrifying destructive power of the atomic bomb.

Beginning in 1934, Italian physicist Enrico Fermi and his colleagues fired neutrons at a variety of stable elements and discovered that they could produce new radioactive ones.

Lise Meitner and Otto Hahn, in Berlin, began bombarding uranium and other elements with neutrons and discovered what appeared to be barium, a lighter element, among the decay products of the uranium bombardment. Most scientists thought that hitting a large nucleus like uranium might result in small changes in the number of neutrons or protons, but no one expected the nucleus to split apart.

Meitner calculated that the combined mass of the new nuclei was slightly less than that of the original uranium nucleus. The difference was being converted into energy, in accord with Albert Einstein's

Enrico Fermi discovered that radioactive elements could be produced by bombarding stable elements with neutrons.

Chicago Pile-1 – the first nuclear reactor, built in an abandoned squash court at the University of Chicago.

famous $E=mc^2$ equation. This was the discovery of nuclear fission that would lead to atomic energy and the atomic bomb.

Fermi believed that neutrons might be emitted in the fission process, leading to the possibility of a chain reaction, and began work on possible designs for a uranium chain reactor. In December 1941, in an abandoned squash court on the University of Chicago campus, workers and scientists, supervised by Fermi, worked under freezing conditions to build a nuclear reactor. Construction was finished a year later and on 2 December 1942 the pile went critical. The world's first self-sustaining nuclear chain reaction was in progress. The

The detonation of the first atomic bomb.

experiment proved that nuclear energy could generate power, even if in this case it was just a modest 200 watts.

One of the products of the nuclear reaction was plutonium, which was first made in December 1940 by bombarding uranium with alpha particles. Within a couple of months element 94 had been added to the periodic table of the elements, its basic chemistry shown to be like that of uranium. To begin with, the amounts of plutonium produced were tiny; by August 1942 there was only 3 millionths of a gram. It was determined that plutonium would undergo fission with a devastating release of energy.

BUILDING THE BOMB

Acquiring the highly radioactive plutonium in sufficient quantities meant constructing new reactors to increase production. At the same time work was going on to design an atomic bomb. Detonating an atomic weapon requires a critical mass of fissionable material, which means enough uranium or plutonium to produce a chain reaction. The more fissionable material, the greater the odds that the chain reaction will occur.

At Los Alamos, New Mexico under the direction of theoretical physicist J. Robert Oppenheimer, methods were developed of turning the products of the reactors into pure metal and fabricating it into the shapes necessary for the bomb. After three years of cutting-edge research and experimentation the world's first plutonium weapon, dubbed the 'Gadget', was ready.

On 16 July 1945 at the Trinity test site the Gadget exploded over the New Mexico desert, vaporizing the firing tower. A huge blast and heat wave burned out across the desert along with a sleet of invisible radiation. A 180,000 kg (200 ton) steel container standing almost 1 kilometre (0.6 miles) from ground zero was pushed aside by the blast. An orange and yellow fireball spread upwards while a second column rose and flattened into the mushroom shape that became the symbol of the destructive power of the atomic age. Famously, Robert Oppenheimer quoted a line from the *Bhagavad Gita*: 'Now I am become Death, the destroyer of worlds.' The only two nuclear weapons ever used in armed conflict to date were detonated in 1945 over the Japanese cities of Hiroshima, on 6 August, and Nagasaki, on 9 August, with devastating loss of life.

NUCLEAR ENERGY

The chain reaction in an atomic weapon is designed to increase in intensity so that the fission process takes place as rapidly as possible, producing energy in a single catastrophic burst. In a nuclear reactor the chain reaction has to be controlled. When a reactor is started up, neutron-absorbing control rods are removed from the core to initiate the chain reaction. When the reactor reaches the desired power level the control rods are partially reinserted to limit the quantity of neutrons produced. Fully inserting the control rods shuts down the reaction altogether.

The first nuclear reactor to produce electricity started operating in December 1951 in Idaho. It produced just enough electricity to power four light bulbs. Today, around 10 per cent of the world's electricity is produced by nuclear power although there are moves to increase this as part of the struggle to combat global warming. Just like in fossil fuel power stations, heat, in this case generated by nuclear fission, is used to heat water and make steam. The steam powers turbines and the turbines spin generators

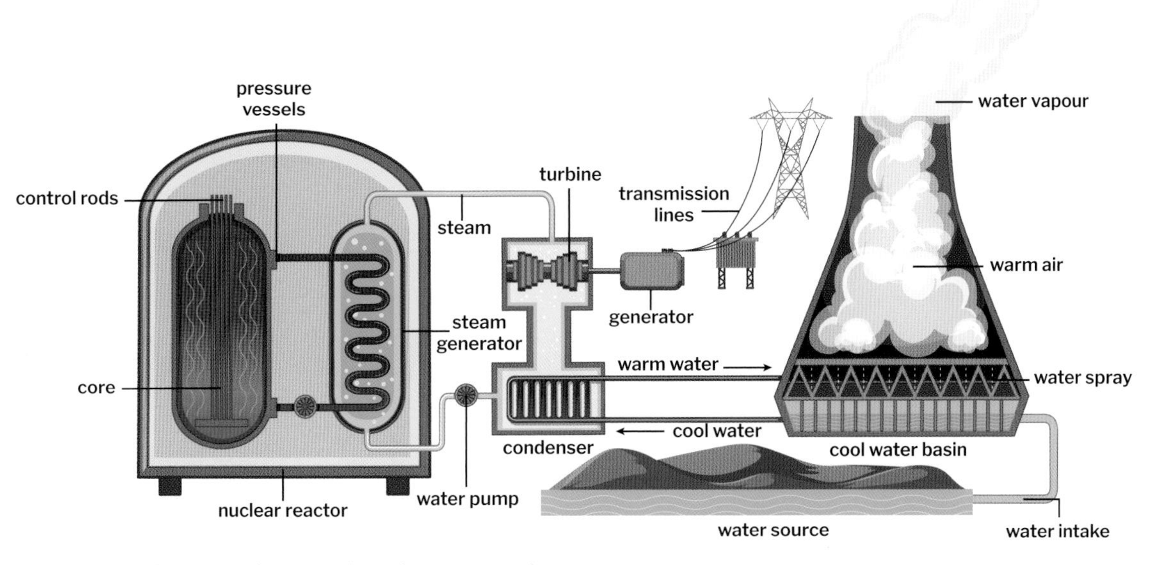

A schematic of a typical nuclear power plant.

to produce electricity.

One of the main types of reactor in use today is the pressurized water reactor (PWR). Other types of reactor include thermal reactors, operating at temperatures in excess of 1,000°C (1,832°F). In the 'pebble bed' reactor the nuclear fuel is encased in tennis ball-sized graphite spheres designed so that, as the fuel temperature rises, the reactor's power output reduces, ensuring it will not overheat. Fast reactors use fuel with much higher concentrations of fissile material and depend on fast neutrons to maintain nuclear reactions. They can be used to produce extra fuel (called breeder reactors).

Nuclear energy has always had its opponents, concerned about the dangers of radiation and the disposal of hazardous waste. A number of unfortunate incidents involving nuclear plants over the years have done little to allay fears. In April 1986 a serious incident occurred when one of four nuclear reactors at the Chernobyl power station in Ukraine exploded. Tens of thousands fled from their homes and many lost their lives then and in subsequent years as a result of radiation. In March 2011 in Fukushima, Japan, an earthquake and subsequent tsunami struck the northeast coast of Japan, triggering a catastrophic meltdown at the Fukushima Daiichi nuclear power plant.

35

Annihilating opposites

Antimatter

Antimatter drives are a popular means of powering science fiction starships across space. But antimatter itself is real and has even been made here on earth. All matter has an antimatter equivalent so why is the universe almost entirely filled with matter? What happened to all the antimatter?

British physicist Paul Dirac first proposed the existence of antimatter in 1928, when he produced an equation that married together Ernst Schrödinger's quantum mechanics wavefunction equation with Albert Einstein's special relativity $E=mc^2$ equation to describe the movement of electrons. One consequence of Dirac's equation was that mathematically it was equally valid for there to be a particle that was the exact opposite of an electron – positively charged where the electron was negatively charged. Dirac called this an antielectron.

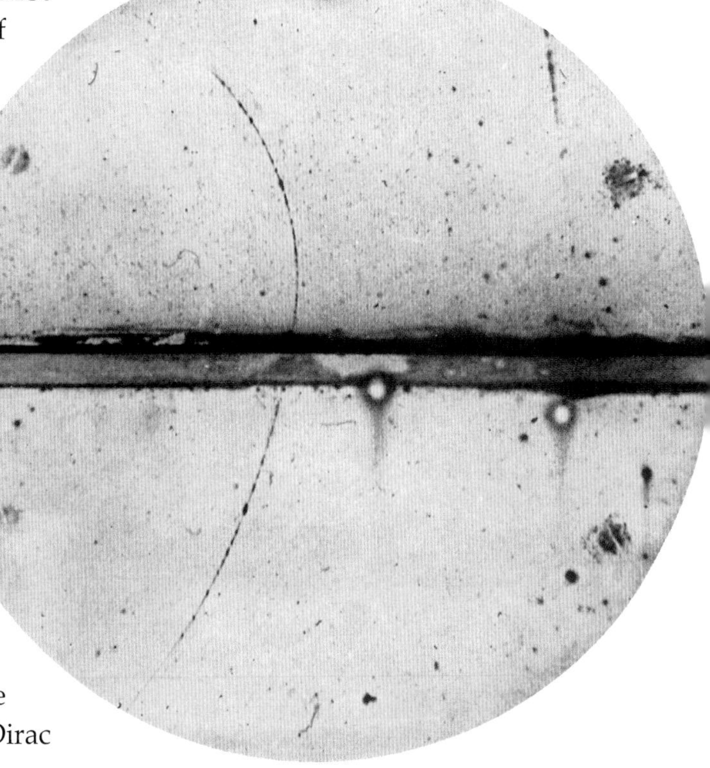

Carl Anderson confirmed the existence of the antielectron, or positron, when he captured its curving track in this cloud chamber image.

In 1932, American physicist Carl Anderson confirmed the existence of the antielectron when he was studying cosmic rays in a cloud chamber at the California Institute of Technology. He spotted a particle with the same mass as an electron, but positively charged. He called the antielectron a 'positron', concluding that the impact of the cosmic rays had produced an electron / positron pair of particles.

ANTIMATTER ASYMMETRY

Since the discovery in 1932 of the positron, or antielectron, it has been confirmed that all matter has an antimatter equivalent. Antimatter particles have the same mass as their matter counterparts but are opposite in characteristics such as electric charge. Research confirms that matter and antimatter particles are always produced in pairs. If particle and antiparticle meet they annihilate each other.

According to theory, equal amounts of matter and antimatter should have been created in the Big Bang that started off the universe, yet today everything we see is composed almost entirely of matter. Where did all the antimatter go? One of physics' greatest challenges is to explain the asymmetry between matter and antimatter in the universe. A possible solution involves violating what is thought to be a fundamental symmetry in nature.

CP symmetry has two components: charge conjugation (C) and parity (P). Charge conjugation transforms a particle into its corresponding antiparticle, mapping matter into antimatter. According to charge conjugation symmetry the laws of physics apply equally to particle and antiparticle. Parity reverses the space coordinates. Applying P to an electron moving with a velocity v from left to right, flips its direction so it is now moving with a velocity -v, from right to left. Parity conservation means that the mirror images of a reaction occur at the same rate – if particles are being emitted up and to the right an equal number should be emitted down and to the left. Applying CP to matter gives us the corresponding antimatter mirror image.

In 1964, physicists James Cronin and Val Fitch made the surprising discovery that particles called neutral kaons, which were formed from a strange quark and a down antiquark, did not obey the CP symmetry rules. The neutral kaons could transform into their antiparticles (in which each quark is replaced by its opposite) but with different probabilities of each happening. The difference was small, just one in a thousand, but it was enough to demonstrate a difference between matter and antimatter. Symmetry had been violated.

In 1972, Japanese theoretical physicists Makoto Kobayashi and Toshihide Maskawa brought CP violation into the Standard Model by proposing the existence of six types of quark. With six quarks, quantum mixing would allow for the occurrence of very rare CP-violating decays. Their predictions were borne out by the discovery of the

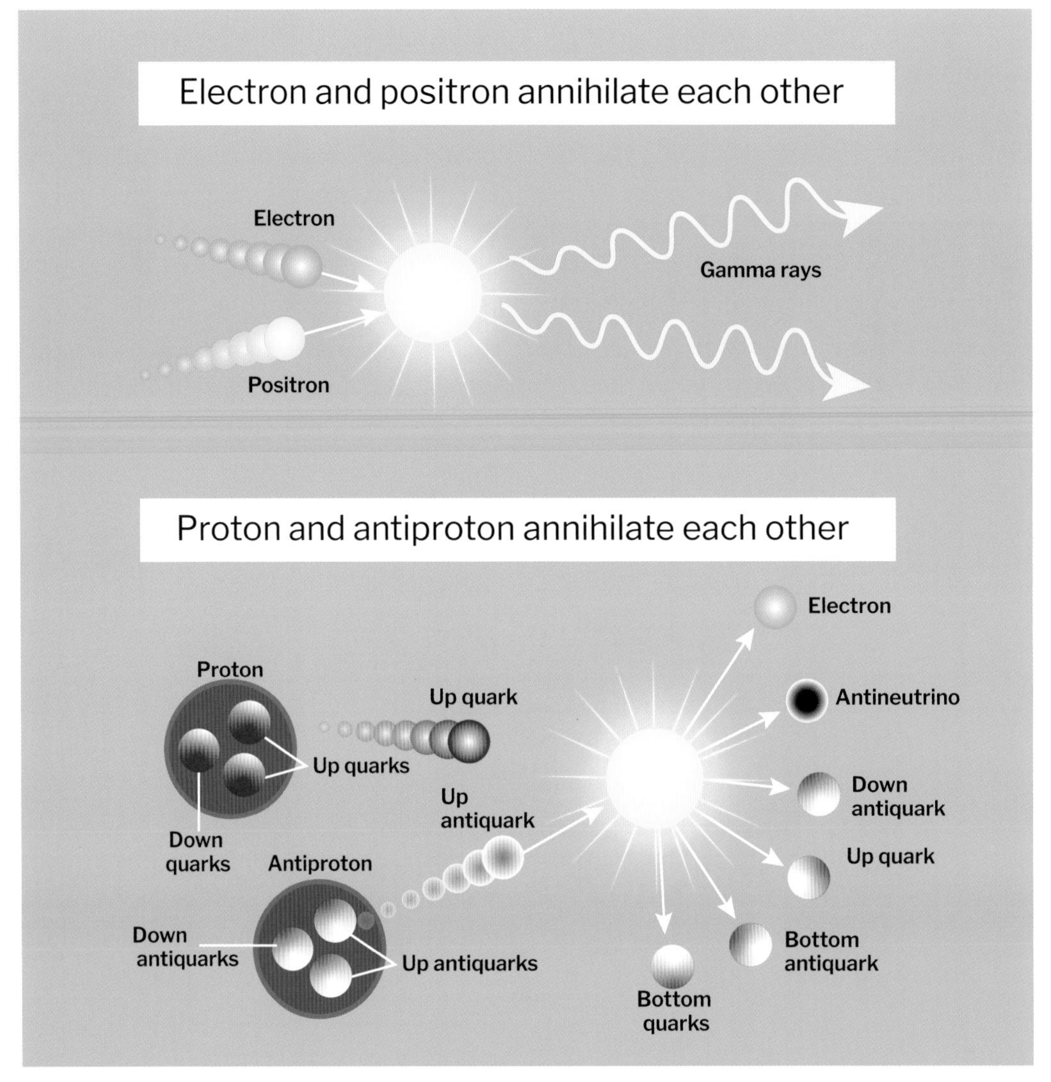

An electron and positron annihilate in a burst of gamma rays. Proton and antiproton annihilation produces a shower of quarks and other particles.

bottom and top quarks, in 1977 and 1995, respectively. Unfortunately, this theory is still unable to provide a full explanation for the amount of matter in the universe as its predictions are still several orders of magnitude short of what is observed. Evidently there is some as yet hidden process at work that tips the odds in favour of matter over antimatter, but just what that process might be remains elusive.

36

Completing the picture

The Higgs boson

The Standard Model answered many questions, providing an explanation for the workings of particle physics based around a few fundamental forces and a scattering of fundamental particles. But it didn't provide an answer for everything. One outstanding question was why some particles which theory said should be massless were actually massive.

In 1963, Sheldon Glashow, Abdus Salam and Steven Weinberg suggested that the weak nuclear force and the electromagnetic force could be combined in what would be called the electroweak force. They predicted that this would occur at energy and temperature levels similar to those which were found shortly after the Big Bang, when the universe expanded rapidly from a super-dense subatomic state. In 1983, physicists at CERN, the European Laboratory for Particle Physics, achieved these temperatures in a particle accelerator and showed that the electromagnetic force and weak nuclear force were indeed related.

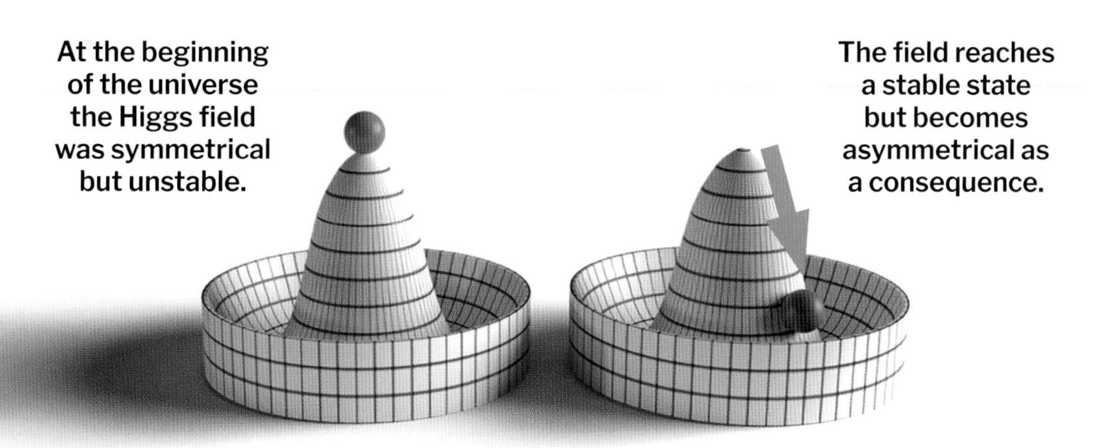

The collapse of the Higgs field shortly after the Big Bang is thought to have given mass to the W and Z bosons which carry the weak nuclear force.

Peter Higgs, whose research added greatly to our understanding of why mass exists in the universe.

However, the equations that described the unified force required the force-carrying particles to be massless. While this was certainly true of the photon, scientists knew that the W and Z bosons, responsible for carrying the weak force, would have to be heavy, nearly 100 times more massive than a proton in fact, to account for their short range. Theorists Robert Brout, François Englert and Peter Higgs proposed a solution to the problem by suggesting that what is now called the Brout-Englert-Higgs mechanism gives mass to the W and Z particles through their interaction with the 'Higgs field', which is believed to permeate the entire universe.

The more a particle interacts with the Higgs field the more mass it has, and the heavier it is. As the universe cooled and expanded following the Big Bang the Higgs field grew with it. Particles like the photon do not interact with it and have no mass at all whereas particles like the W and Z bosons interact strongly with it and are consequently massive and slow-moving. It is only because particles gain mass through interacting with the Higgs field that the formation of stars, planets and life becomes possible. In common with all fundamental fields, the Higgs field has an associated force-carrying particle – the Higgs boson. The problem facing physics now was to find it.

The search for the Higgs boson took on such significance that it gained the nickname of 'the God particle' much to the annoyance of Higgs and his colleagues. Little progress was made until CERN's Large Hadron Collider, the world's most

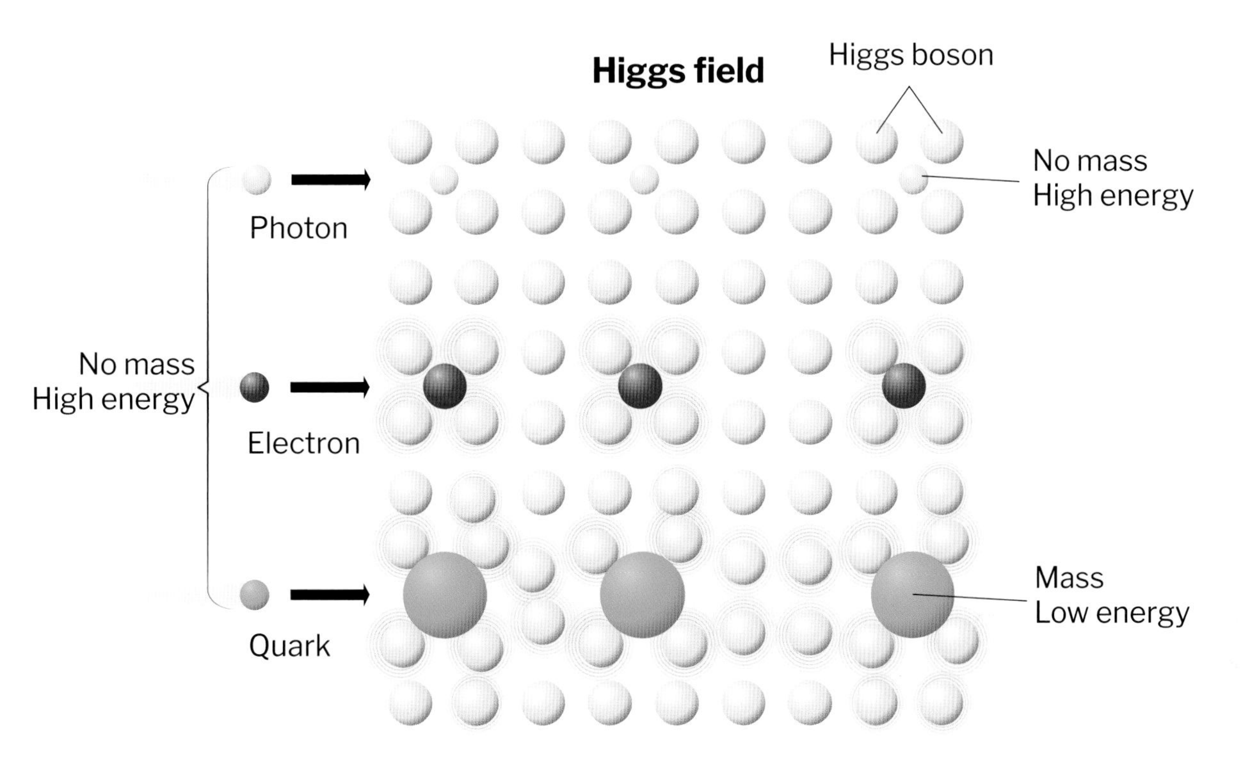

The more a particle interacts with the Higgs field, the greater its mass. The massless photon doesn't interact at all.

powerful particle accelerator, came online in 2008. Researchers at CERN combed through trillions of particle collisions looking for evidence of the Higgs boson. Finally, on 4 July 2012, they announced that they had observed a new particle of a mass that was consistent with the predicted value for the Higgs boson. Two teams worked separately, not discussing their results with each other, to ensure that the findings were genuine. The particle that had been discovered was short-lived, had no electrical charge and decayed in ways that theory predicted that the Higgs boson should. It also appeared to have zero spin, a feature peculiar to the Higgs boson.

Since its discovery physicists have continued to study how strongly the Higgs boson interacts with other particles, to see if this matches theoretical predictions. Questions such as whether the Higgs is a one-of-a-kind particle or representative of a whole new class of particles remain to be answered. How does the Higgs boson itself acquire its mass? Does it interact with itself somehow or is there yet another mechanism waiting to be discovered? What role did the Higgs boson play in the formation of the universe? Can it, for example, explain the imbalance between matter and antimatter? There is still much to learn about the mysterious Higgs.

37

It isn't the same for everyone

Relativity from Galileo to Einstein

Until the beginning of the 20th century physics was mostly concerned with refining a mechanistic view of the universe. Space and time were the unchanging backdrop against which the events of interest to physicists took place. Albert Einstein's relativity theories would upend that perspective.

The roots of relativity stretch back to the 17th century and the work of another great physicist, Italian mathematician and scientist Galileo Galilei. As part of his influential work on moving objects, Galileo explored the idea that all motion is relative and that it only makes sense to speak of movement in relation to something else.

Galileo imagined a passenger aboard a ship sailing across a perfectly smooth lake. If the ship continues to move at constant speed and direction the passenger will not feel its motion, just as a modern-day traveller doesn't notice the motion of an aircraft cruising through the sky. No unit of measurement is absolute – we can only make measurements with reference to something else. To say something is moving only has meaning if we can say what it is moving relative to. If a fellow passenger on an express train tosses you an apple it comes towards you at a catchable speed, but to a trainspotter by the trackside, passengers, apple and train flash by at 100 km/h (62 mph).

Galileo asked, is there any way in which the passenger can determine that the ship is moving without going on deck and looking? He concluded that there wasn't. Any mechanical experiment performed inside the ship, always provided that it was moving with constant speed in a constant direction, would give exactly the same results as a similar experiment carried out on shore. From these observations Galileo put forward his own relativity hypothesis:

> *Any two observers moving at constant speed and direction with respect to one another will obtain the same results for all mechanical experiments.*

FRAMES OF REFERENCE

Galileo's assertion was that the laws of physics apply equally for all freely

Why does a ball appear to be simply going up and down? (While looking from inside a moving train)

Why does a ball appear to be following a parabolic path? (While looking from outside a train)

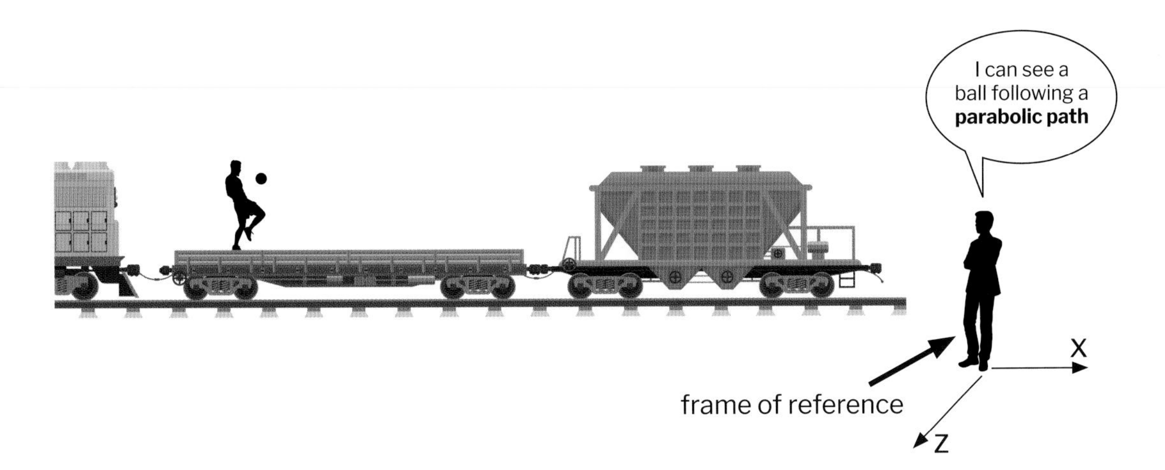

The path followed by a moving object can appear differently depending on the observer's frame of reference.

moving observers, whatever their speed of motion. Isaac Newton later pointed out in his first law of motion that an inertial state, moving with a constant speed and direction, is the default for any object not being acted on by a force. Inertial motion is simply motion at a uniform speed in a straight line. Objects in uniform motion, moving at a constant speed and direction relative to one another, are said to share an inertial frame of reference.

The motion of an object appears different according to your frame of reference.

Newton's idea of absolute motion was generally accepted. This was the idea that an object could be said to be moving without reference to anything else; an idea that required that there must also be a state of absolute rest. Either it was moving, or it wasn't. Newton wrote: 'Absolute motion is the translation of a body from one absolute place into another; and relative motion, the translation from one relative place into another.'

SPECIAL RELATIVITY

The idea that motion has no meaning without a frame of reference is central to Einstein's theory of special relativity. The theory is special in that it deals with the special case of objects moving at constant velocity relative to each other in an inertial frame of reference. Einstein set out his principle of relativity in a paper published in 1905 called 'On the Electrodynamics of Moving Bodies'. It took as its starting point something that had been known to physicists for almost a hundred years – an electric current is generated whether a magnet is moved inside a coil of wire, or if the magnet remains fixed and the coil is moved. The general assumption was that two different mechanisms were at work – one where the moving magnet produces the current and another where the current is produced by the moving coil. The distinction between the moving magnet and the moving coil depended on the view still held by most scientists that, in accordance with Newton, there was such a thing as a state of absolute rest.

Einstein rejected this notion, declaring that the idea of absolute rest was flawed and unnecessary. It didn't matter whether it was the magnet or the coil that was moving, it was their movement relative to each other that generated the current. Einstein wrote:

'The same laws of electrodynamics and optics will be valid for all frames of reference for which the laws of mechanics hold good.'

Another way of saying this is that the laws of physics are the same in all inertial frames of reference. No matter how fast you travel, or in which direction, the laws remain the same, which means that any experiment carried out will produce results that are in accordance with the laws. In many ways this was

Einstein realized that a moving magnet producing an electric current, and a moving current producing a magnetic field, were two aspects of the same phenomenon. It was the relative motion of one to the other that was important.

what Galileo was saying back in 1632. No experiment can determine the motion of the observer in an inertial frame.

Einstein's theory of special relativity did away with the notion of absolute rest and absolute motion. On one, perhaps apocryphal occasion, Einstein supposedly asked a bemused ticket inspector, 'Does Oxford stop at this train?'

38

The constancy of light

The absolute nature of light speed

Albert Einstein asked the question: 'Does light behave the same way as everything else? Is the speed of light dependent on the motion of the observer?' This brought Einstein to the idea upon which he founded his theory of special relativity: that the speed of light is a constant. Some things may be relative, but the speed of light is absolute.

James Clerk Maxwell determined that the speed of an electromagnetic wave is not measured relative to anything else but is a constant defined by the properties of the vacuum of space through which the wave moves. The speed of light, 299,792,458 metres per second (186,282 miles per second), is determined by the very nature of the universe.

Because Maxwell's equations hold true in any inertial frame, two observers moving relative to each other, each measuring the speed of a beam of light relative to themselves, will both get the same answer – even if one is moving in the same direction as the beam of light and one away from it, in direct contradiction to Newton's absolutist standpoint. Once Einstein had adopted his principle of relativity, he realized that it was impossible for both Newton and Maxwell to be right.

The theory of special relativity derives from the simple fact that the speed of light is a constant for all observers and is independent of the velocity of the object emitting the light. This made little sense in Newtonian terms, in which speeds are added together. For example, a fast bowler adds the speed of his run to the

Light speed defined

In 1983, the General Conference on Weights and Measures officially defined the speed of light to be:

c = 299,792,458 m/s.

Scientists use 'c' as the symbol for the speed of light from the Latin word 'celeritas', which means 'swiftness'.

At the same time the metre came to be defined as the distance light travels in one 299,792,458th of a second.

The Michelson–Morley experiment

Waves need some sort of medium to carry them – sound waves travel to your ear through the medium of the air, for example. It was the view of 19th-century scientists that electromagnetic waves must also travel through a medium to cross space. They called it 'ether'.

American scientists Albert Michelson and Edward Morley carried out a series of precise experiments aimed at measuring the effects of the ether on the light that passed through it. Their crucial experiment, carried out in 1887, set out to measure the speed of light in different directions and so determine the speed of the ether relative to the earth.

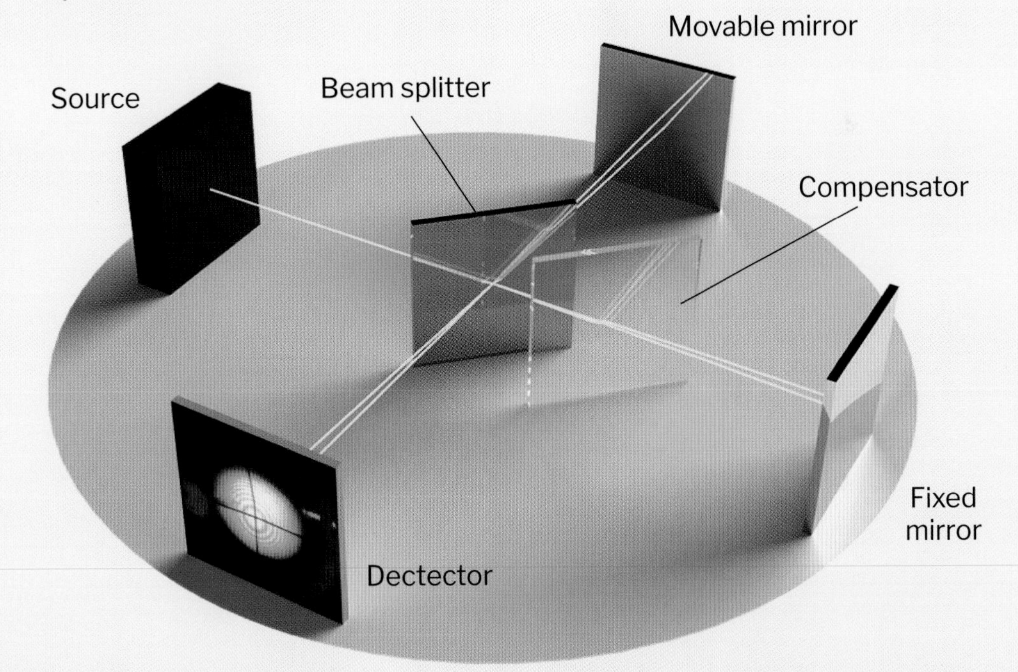

Michelson and Morley's experiment demonstrated that the ether didn't exist and the speed of light remained constant.

Michelson and Morley discovered that it made not one bit of difference which direction or what time of day they took their measurements from. The speed of the light beams was always the same.

Physicists, including Michelson himself, were perplexed by this result and looked for ways to reconcile these findings with their belief in the ether. One suggestion seemed little better than a fudge. Working independently of each other, Dutch physicist Hendrik Lorentz and Irish physicist George FitzGerald came up with the same solution to the problem. In 1889, FitzGerald published a short paper in which he proposed that the results of the Michelson–Morley experiment could be explained if objects were reduced in length as they travelled through the ether. Lorentz put forward an almost identical proposal in 1892.

The reduction in length due to the Lorentz–FitzGerald contraction, as it came to be called, was infinitesimal, amounting to just a couple of centimetres for an object the size of the earth, but it would be enough to explain Michelson and Morley's results. Einstein would show that the phenomenon was real, but not for the reasons Lorentz and FitzGerald suggested.

$v = 0$, $L^* = L$

$v = 0.87c$, $L^* = 0.5L$

$v = 0.995c$, $L^* = 0.25L$

$v = 0.999c$, $L^* = 0.045L$

$v \to c$, $L^* \to 0$

The closer an object approaches the speed of light the more it contracts in the direction of travel.

speed at which the ball is released from his hand. But a beam of light projected from an orbiting spacecraft travels at the same speed as one projected from the (relatively) stationary surface of the planet below it.

A car travelling at 80 km/h (49.7 mph) overtaking another car travelling at 70 km/h (43.5 mph) passes the second car at 10 km/h (6.2 mph). That's simple physics. Now imagine sending a laser signal to a spacecraft that's heading away from you at half the speed of light. Common sense says that the laser light should reach the spacecraft at half-light speed because it has to catch up with the spacecraft, but common sense is wrong. The beam will still arrive at the spacecraft at approximately 300,000 km/s (186,000 miles/sec).

Newtonian physics tells us that velocity equals distance travelled divided by the time taken to cover that distance – $v = d/t$. So, if the speed of light, v, always remains the same, whatever the other two values, then it follows that d and t, distance, or space, and time, must change. For you and the pilot in the spacecraft to agree on the speed the beam reaches the spacecraft you have to agree on the time it takes to get there. Since the speed of light remains constant it follows that our notions of space and time have to change. It was from this fundamental rethink that the radical ideas of special relativity would flow.

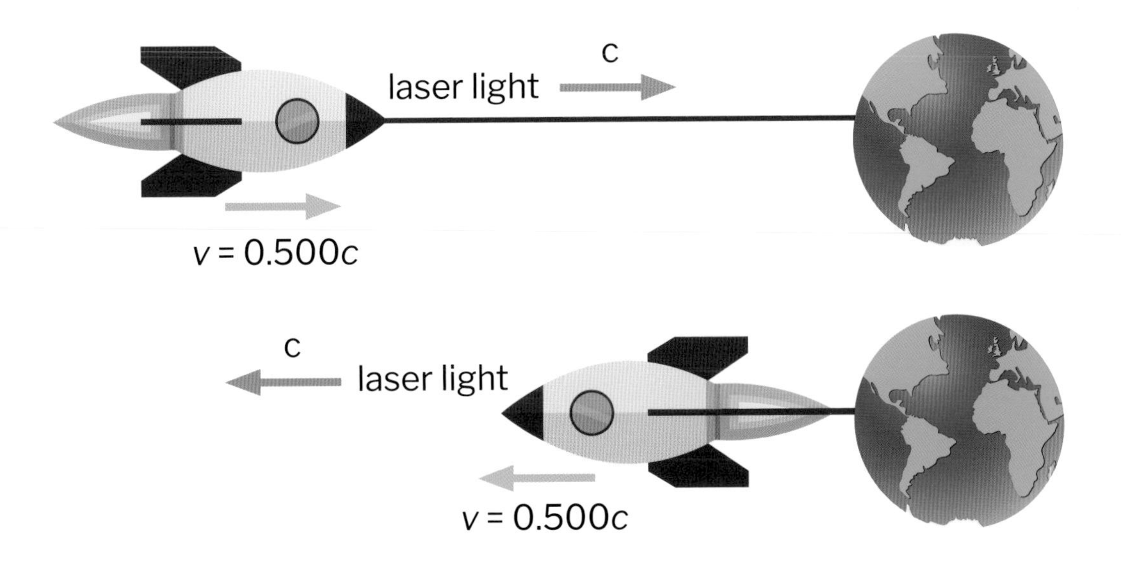

No matter which direction the spacecraft is travelling the speed of the light beam will always be the same.

39

Adventures in time and space

The theory of special relativity

Isaac Newton wrote that, 'Time exists in and of itself and flows equably without reference to anything external.' In Newton's view, time always ticked by at the same steady pace. Ten seconds for me will be ten seconds for you too, no matter what we might be doing when we measure it. Einstein declared, however, that time is relative and passes differently in all moving frames of reference.

Einstein argued that Newton's absolute time and absolute space should be replaced by absolute spacetime. The mathematics of relativity demonstrate that space and time are inextricably linked, and both are altered when we approach near-light speeds. Only by considering space and time together can we give an accurate description of what is observed at light speed. Motion, distance and the duration of time, all take place against the unchanging backdrop of spacetime, the geometry of which is rigidly dictated by the speed of light. Absolute spacetime is as crucial to the understanding of special relativity as the absolute time and absolute space it replaced were to Newtonian physics.

THE END OF SIMULTANEITY

One effect of special relativity that Einstein thought particularly important is the relativity of simultaneity. Two events that appeared to one observer to have occurred simultaneously might not appear to have done so to a second observer moving relative to the first. According to Einstein, there is no way in which we can say one observer is correct and the other one wrong. They are, in fact, both right!

Einstein explained the conundrum in terms of a thought experiment. Imagine that you are watching a thunderstorm and suddenly two buildings that you know to be equidistant from you are struck by lightning. You would say that the two had been struck simultaneously. However, for another observer moving towards one building and away from the other the light from the strike on the second building will take longer to arrive than the light from the one the observer is moving towards, so the strikes will not appear simultaneously.

The principle of relativity says that

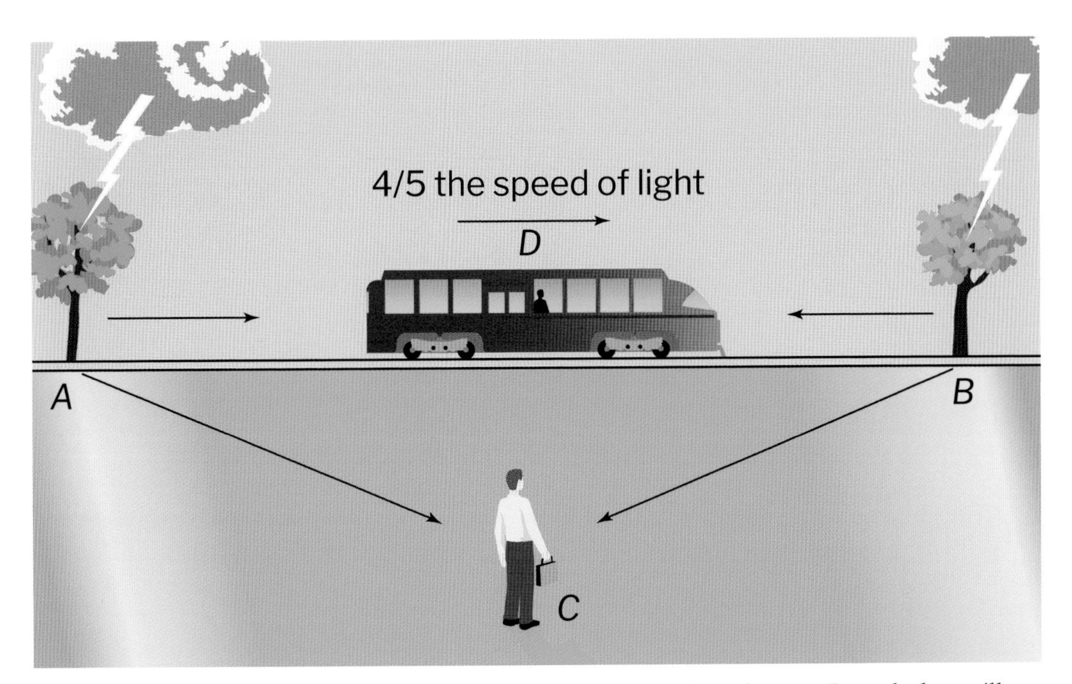

For observer C lightning strikes trees A and B simultaneously, but observer D on the bus will see tree B struck first.

there is no way to insist that you are at rest and the other observer is in motion. You are simply in motion relative to each other. There is, therefore, no 'right' answer as to whether or not the lightning strikes occur simultaneously.

The end of simultaneity is another nail in the coffin of absolute time. Two observers in relative motion will have clocks that tick at different rates; the effect becomes more noticeable approaching the speed of light, but it is there, if infinitesimally, at low relative speeds too. Time passes differently for all moving reference frames.

EINSTEIN'S MIRROR

Einstein asked himself a deceptively simple question: If I held a mirror while travelling at close to the speed of light, would I see my reflection? How would light reach the mirror if the mirror was travelling at light speed? It was thought experiments like this that allowed Einstein to lay the foundations of relativity. If the speed of light is a constant, then, no matter how fast he is going, the light going from Einstein to the mirror and back again will always be travelling at 300,000 km/s (186,000 miles/sec), because the speed of light does not change. To make the calculation work, not only does time have to be slowed down, but the distance travelled by the light beam has to decrease as well.

According to special relativity, the faster you travel through space, the slower you travel through time. Approaching

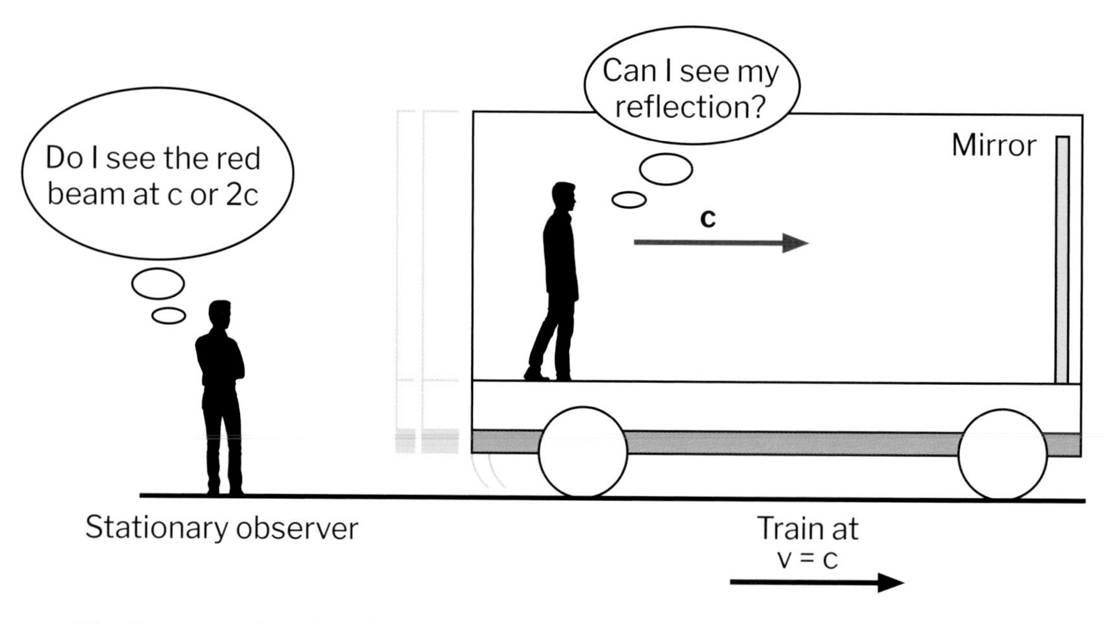

The observer on the train will see their reflection because they are in an inertial frame with no way to determine their velocity.

the speed of light the intervals between events lengthen, so time seems to slow down. This phenomenon is called time dilation. If an object could achieve the speed of light, then time would appear to stop entirely. Scientists at CERN's Large Hadron Collider, where particles are smashed together at near-light-speed velocities, have to take the effects of time dilation into account when interpreting the results of their experiments.

THE LORENTZ–FITZGERALD CONTRACTION REVISITED

Another of the strange consequences of the speed of light remaining constant to all observers is that a moving object appears to shrink along the direction of motion. At the speed of light, the length of the object would be zero. This phenomenon is called the Lorentz–FitzGerald contraction after the two physicists who proposed it in 1889 and 1892 as a solution to the failure of the Michelson–Morley experiment. It was Einstein who showed the phenomenon was real but a consequence of the properties of space and time and not an actual physical compression.

Imagine a spacecraft with mirrors mounted at each end between which a pulse of light bounces. What happens to the pulse as the spacecraft approaches the speed of light?

For a 150-metre-long (500 ft) ship at rest, the return journey for the light beam will take roughly a millionth of a second. However, at 99.5 per cent of the speed of light, time is slowed by around a factor of 10, so the round-trip journey

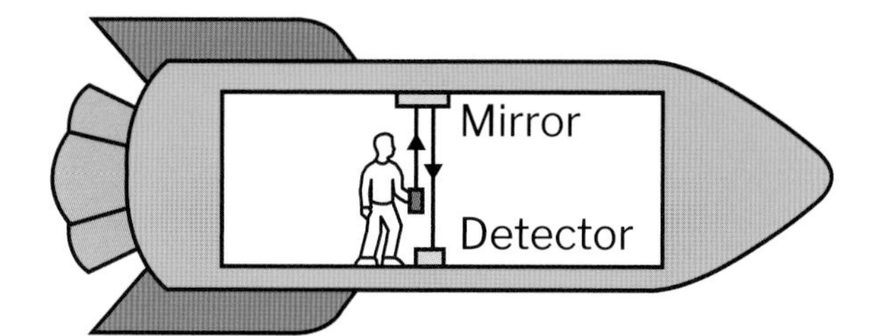

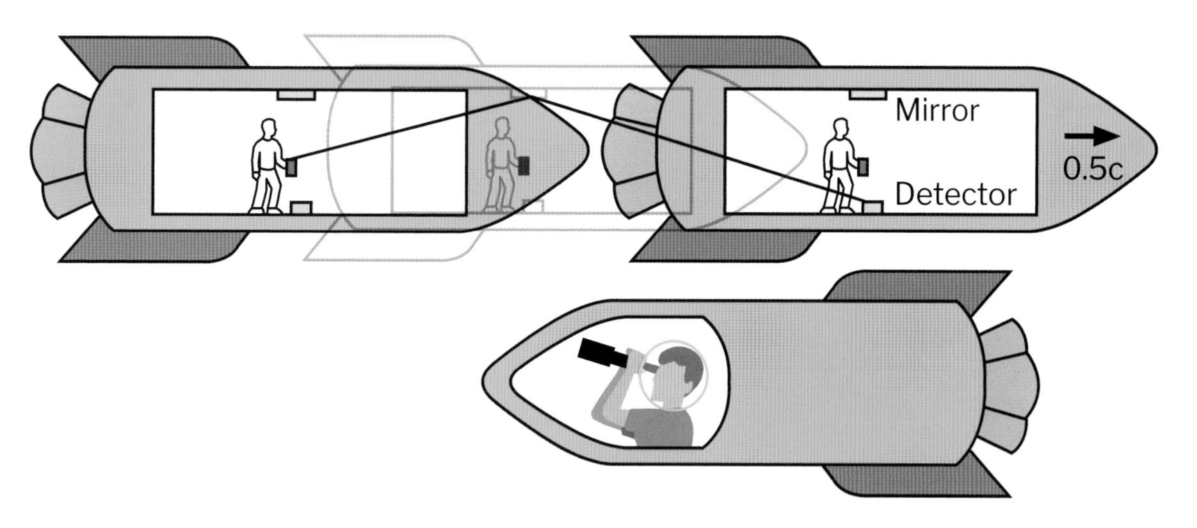

Because the speed of light remains constant our measurements of time and space must change.

time, as measured by an observer, is now just a hundred thousandth of a second. However, the pulse from the back to the front has further to go, because the mirror is retreating from it at close to the speed of light. The return trip from front to back is shorter because the rear mirror is rushing towards the light beam. But no matter whether the mirror is retreating or advancing, the light beam will always reach it at the same speed, roughly 300,000 km/s (186,000 miles/sec), because the speed of light doesn't change. As with Einstein's mirror, time slows down, and the distance travelled by the light decreases.

At roughly 99.5 per cent of light speed, the distance is reduced by a factor of 10 – the same proportion as the time dilation effect. This shrinkage only takes place in the direction of motion and will only be apparent to an observer who is at rest with respect to the moving object. The crew of the spaceship won't perceive any change in their ship's length, or in themselves, rather they would see the observer appear to contract as they streaked by.

A consequence of the contraction of space is to shorten the time it would take to travel to the stars. Imagine a cosmic railway network with tracks stretching from star to star. The faster a spacecraft travels, the shorter the track appears to become, and therefore the shorter the distance to be covered to reach its destination. At 99.5 per cent of light speed the journey to the nearest star would take around five months in ship's time. But for an observer back on earth the trip would appear to take over four years!

In the old Newtonian physics, travel through time and travel through space were held to be two quite separate things. But, according to Einstein, that's not the case. If you are stationary – that is, not moving through space – then all your spacetime movement is through time. When you start to move, some of your movement through time is diverted into movement through space. The speed of your journey through time slows when some of your time-like motion is used for your journey through space. According to special relativity, the combined speed of an object's motion through time and through space is precisely equal to the speed of light. This is an upper speed limit that can't be broken. For an object in motion, time must slow down otherwise the total combined speed through spacetime would exceed the speed of light. At the speed of light all the spacetime movement has become movement through space with nothing left over for movement through time.

$E=mc^2$

One of the most famous equations in all physics is derived from special relativity. Einstein published it as a sort of postscript to the special theory in a short paper, just three pages long, entitled 'Does the Inertia of a Body Depend Upon its Energy Content?' Known sometimes as the law of mass-energy equivalence, $E=mc^2$ effectively says that energy and mass are two aspects of the same thing. If an object gains or loses energy, it loses or gains an equivalent amount of mass in accordance with the formula. For example, the faster an object travels, the greater is its kinetic energy, and the greater also is its mass. The speed of light is a big number – squared it is a very big number indeed. This means that when even a tiny amount of matter is converted into its energy equivalent the yield is colossal, but it also means that there has to be an immense input of energy to see an appreciable increase in mass.

40

The block universe

Mapping spacetime

Einstein argued that Newtonian absolute time and absolute space could be consigned to the dustbin and replaced by absolute spacetime. The mathematics of relativity inextricably link space and time, and both are altered when we approach near-light speeds.

'Henceforth, space by itself, and time by itself, are doomed to fade away into mere shadows, and only a kind of union of the two will preserve an independent reality.' So wrote German mathematician Hermann Minkowski a year after Einstein published the theory of special relativity. In 1907, Minkowski developed a way of visualizing how objects moved through space and time called Minkowski spacetime diagrams.

A Minkowski spacetime diagram uses a coordinate system with time shown vertically on the y axis, and two

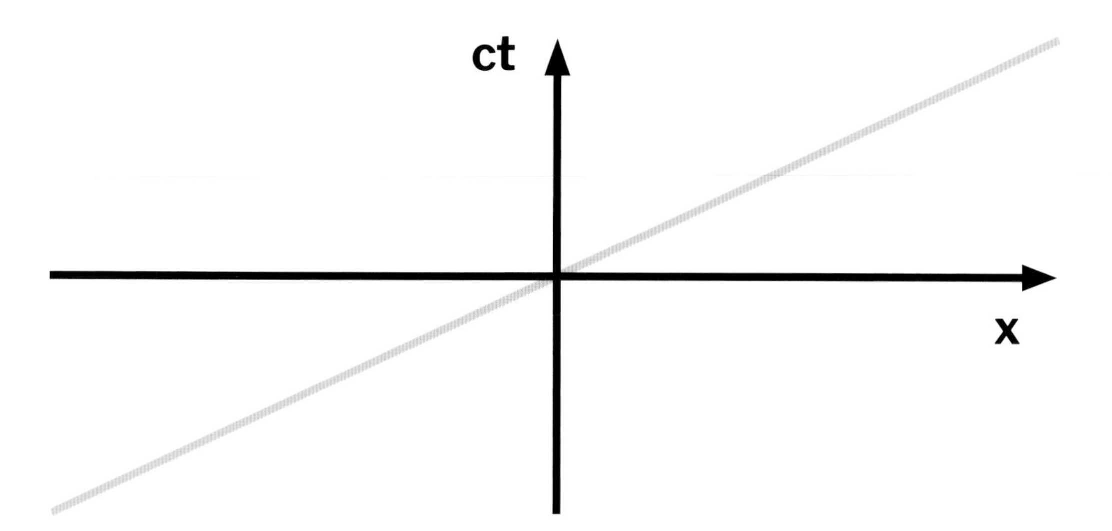

The worldline of a photon makes a 45-degree angle when plotted against space (x) and seconds multiplied by the speed of light (ct).

of the space dimensions represented along the x and z axes like a perspective drawing. An object is not represented as a single point but as a line delineating all the spacetime points at which it exists. This is the object's worldline. If it is in uniform motion, the object's worldline will be straight, but any force acting on it will cause the worldline to curve. The units along the time axis are usually given as seconds multiplied by the speed of light, so that the worldlines of light rays make a 45-degree angle with each axis.

UNIVERSE IN A BOX

To help visualize a path through spacetime, physicists employ a concept called the block universe. Picture the universe as a huge four-dimensional box with time as the fourth dimension. As four dimensions are a little tricky to visualize it's easier to simplify the picture by flattening space into two dimensions and swapping the third spatial dimension, going from left to right, for time. Taking a slice through the box gives us a snapshot of the block

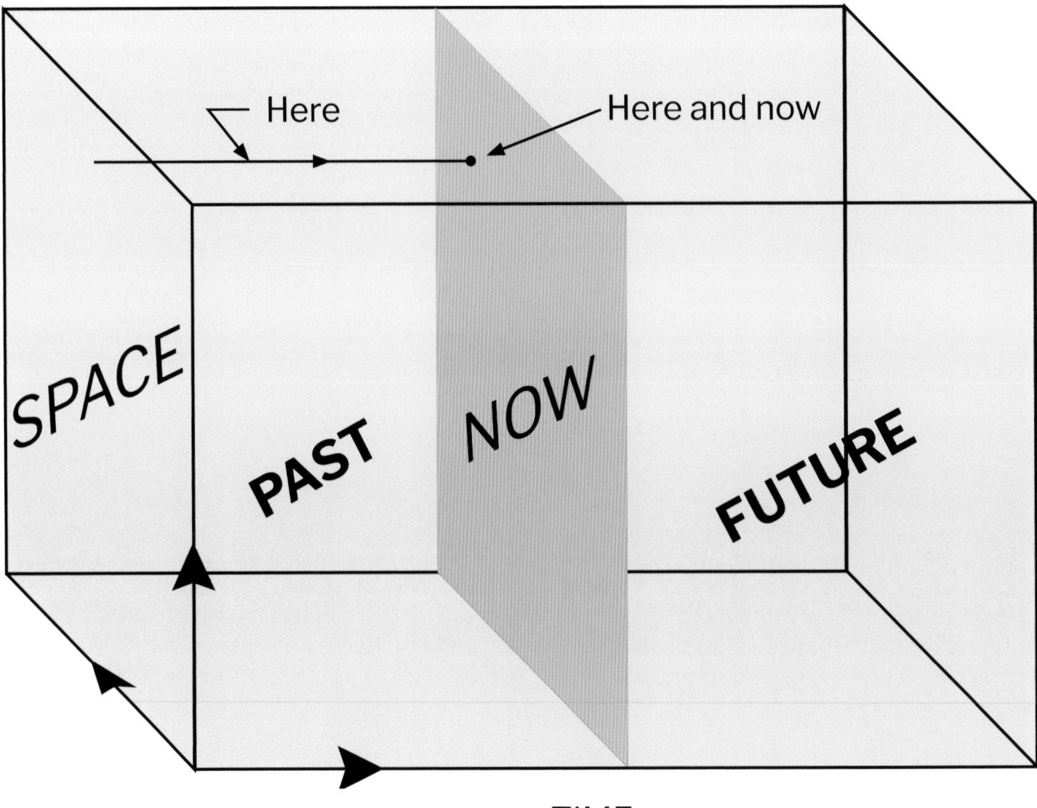

The block universe is an aid to mapping the three dimensions of space and the fourth dimension of time.

universe at a moment in time. Any event at any place in our flat universe can be mapped on the box, its coordinates showing us not just where it happened but also when it happened.

'Now' in the block universe is the slice taken at this very moment. Time 'flows' in a series of infinitesimal jumps from one slice to the next, each jump so small as to be indetectable. In another view, all of the slices exist simultaneously with all past and future events mapped along the timeline, although we are prevented from seeing this due to our inability to step outside the four dimensions of spacetime.

Whether time flickers into the future, like the pages of a flipbook, or whether the future, like the past, already and unalterably exists, Einstein's relativity ties us into a picture of the universe in which space and time are inextricably linked. Observers in relative motion will not be able to agree on which page of the flipbook an event took place. Because of the effects of time dilation and length contraction, spacetime divides into its space part and time part differently for observers in reference frames moving relative to each other. There is no way, for example, to claim unambiguously that an event lasted ten seconds without giving some indication of the reference frame in which the measurement was made.

LIGHT CONES

The fact that nothing can travel faster than light places a restriction on how events can influence each other in spacetime. The path of all possible light-speed worldlines coming from an event spread out from it in a growing circle, like ripples from a pebble tossed into a pond. Imagine all these circles stacking one on top of the other up the timeline to form an inverted cone with its point at the origin of the event. This is called a light cone. The future light cone of the event maps out all the possible future events in spacetime that can be affected by the event. Because nothing can travel faster than light, anything outside the light cone cannot possibly be influenced by, or have any knowledge of, the event.

An exactly symmetrical past light cone also extends out from the event into the past. The past and future light cones divide spacetime into three regions. The absolute future of the event is the region inside the future light cone. It contains everything that can possibly happen as a result of the event. The absolute past of the event is everything inside the past light cone. It contains everything that could possibly have caused or affected the event. Anything outside the past light cone can have had no effect on – or to have caused – the event in question.

The light cone is a map of the geography of spacetime that shows the limits of the possible interactions between events. Everything that lies outside the past and future light cones of an event is 'elsewhere'. Anything in the 'elsewhere' can have no knowledge of the event and can have no influence

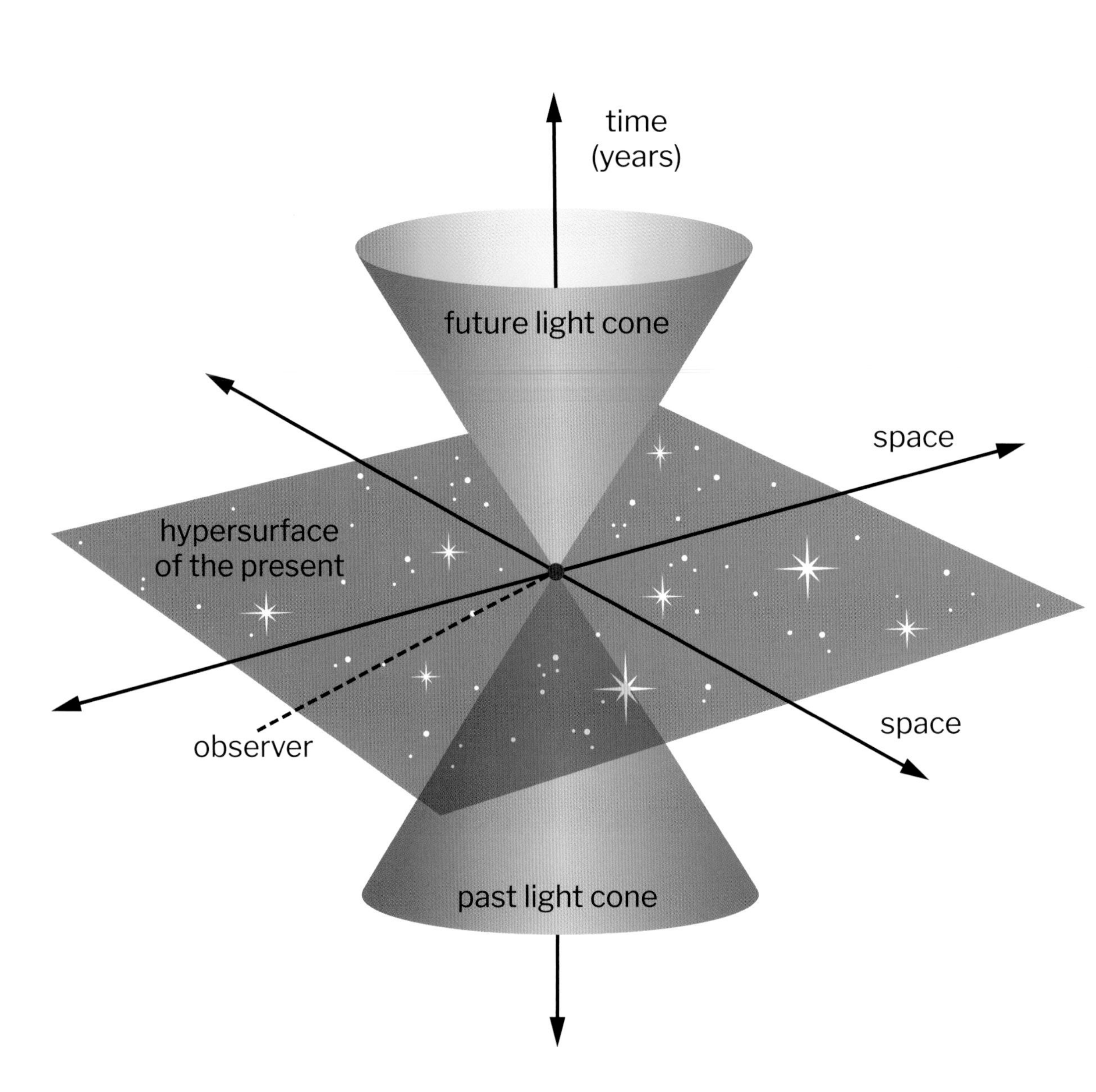

on it or be affected by it. Every event in spacetime has its own light cone so that spacetime is filled with an endless number of infinitely overlapping cones.

The past and future light cones put limits on all the events that can influence an object and all the events it can affect in turn.

41

Einstein's gravity

The theory of general relativity

In setting out the theory of special relativity, Einstein focused solely on objects moving with uniform motion. He chose to ignore objects that were accelerating, and objects affected by gravity. He did this for the very good reason that it made the calculations a lot easier.

Special relativity was built around the fact that light was the fastest thing in the universe. This directly contradicted Newton's ideas about how gravity worked. According to Newton, gravity acted without delay, making its effects felt instantaneously. If the sun were to disappear, the earth, no longer held in place by the sun's gravitational pull, would slingshot out of its orbit in the same second. If Einstein was right, then that couldn't be possible as no influence could reach from the sun to the earth in less time than the eight minutes or so it took light to travel that distance.

So how did gravity make its influence felt? Obviously, it was a force that acted at a distance and didn't require any physical contact to work. Also, unlike any other force, it was impossible to shield yourself from its effects. Newton's law explained how to calculate the effects of gravity but made no attempt to explain what caused it.

FEELING THE FORCE

All observers moving uniformly relative to each other are entitled to say that they are stationary and that it's everyone else who is moving. If you are in uniform motion, then it's impossible to demonstrate that you are actually moving.

Accelerated movement is quite different. We feel a change in speed or direction. Even with your eyes closed you know when the train you're on is going round a curve or slowing down as it enters a station. Acceleration involves inertial forces – the forces that resist a change in speed or direction. These are the forces that push us against the seatbelt when the car brakes suddenly or cause our coffee to part company with its cup when the bus hits a pothole.

In his legendary experiment at the tower of Pisa, Galileo demonstrated that a small stone and a large stone will fall to the ground in the same time. This is because the two stones accelerate towards

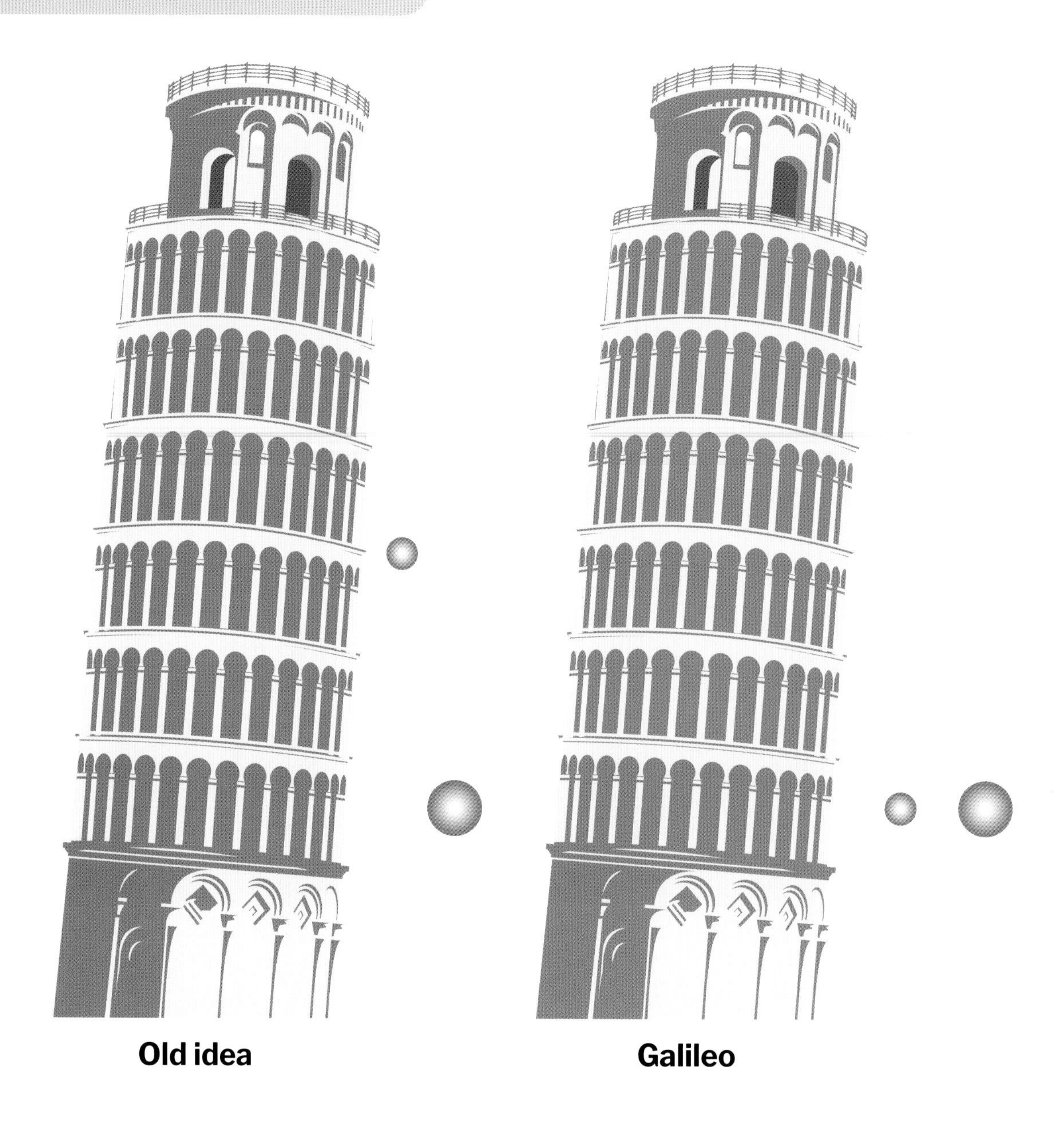

Galileo demonstrated that, contrary to accepted belief at the time, objects of different masses will fall at the same rate.

the ground at the same rate, and this holds true no matter what the difference in their mass. Newton explained this with his second law of motion – force equals mass times acceleration. On earth the acceleration of a falling object due to the force of gravity is approximately 9.8 m/s^2 (32 ft/s^2).

This idea, that gravity accelerates all objects at the same rate regardless

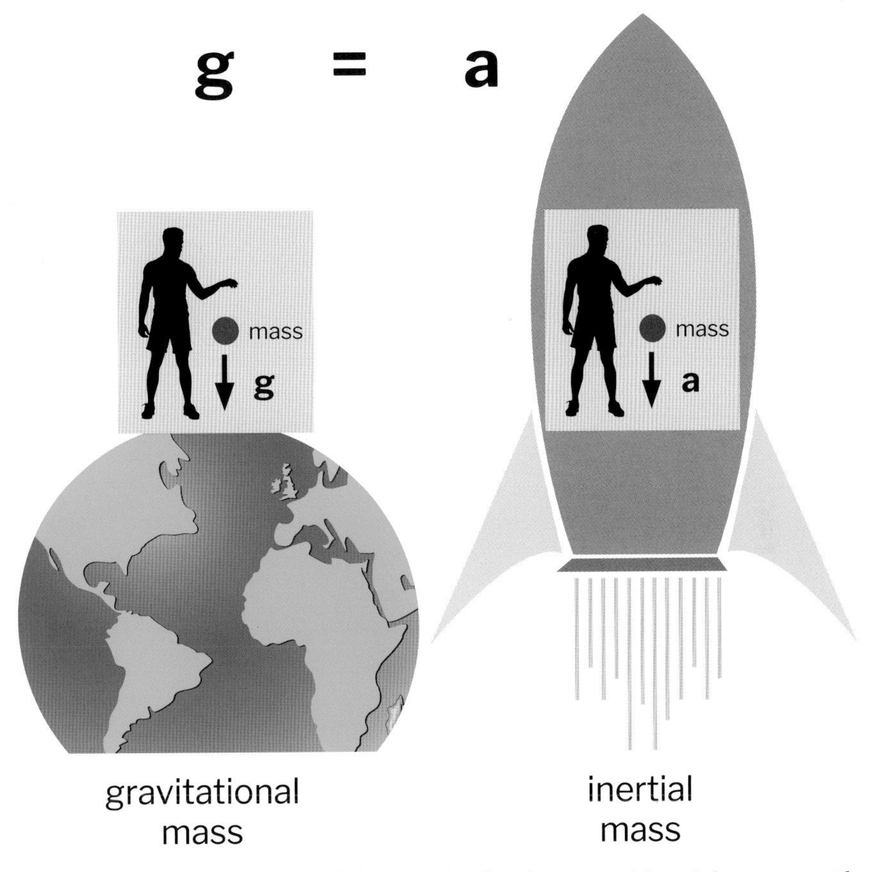

Einstein's equivalence principle determined that gravitational mass and inertial mass were the same and there was no way to distinguish between the effects of gravity and acceleration.

of what they might be made of, is called the 'Universality of Free Fall' or the 'Equivalence Principle'. This is, of course, an ideal situation where gravity is the only force acting and we ignore other forces such as the frictional force of air resistance. Astronaut David Scott convincingly demonstrated the validity of the Equivalence Principle on the surface of the Moon in 1971 (see page 00). When a force is applied to an object its inertial mass can be calculated by measuring its acceleration. The gravitational mass can be calculated by measuring the force of gravity. Both answers will be the same. Einstein was convinced that this couldn't be a coincidence.

HAPPY THOUGHTS

In what he called his 'happiest thought', Einstein formulated a principle of equivalence that stated that the effects of uniform acceleration were

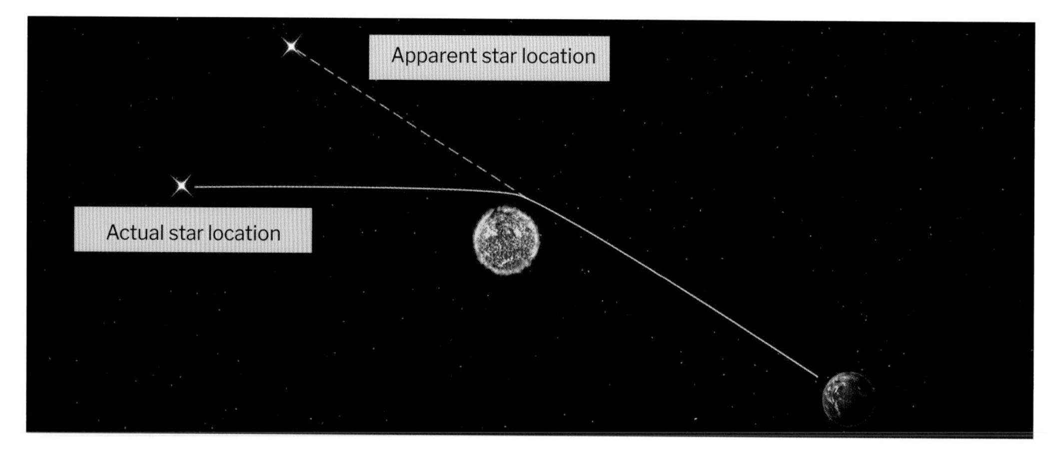

Observations of the bending of starlight during a total eclipse of the sun provided proof that Einstein's ideas were correct.

indistinguishable from the effects of gravity. In another of his famous thought experiments, he imagined someone in a box in deep space under uniform acceleration. An object released in the box would fall to the 'bottom' of the box, that is, in the opposite direction to that in which the box is moving. All objects will fall exactly alike, no matter what their mass or composition, just as if they were in a gravitational field.

According to Einstein's equivalence principle, whether the person in the box was accelerating or not depended on your point of view. The person in the box would believe themselves to be in a gravitational field and not accelerating, but an observer watching the box would see it accelerating uniformly through gravity-free space. Each relative point of view is equally valid. This was what made the inertial mass and the gravitational mass the same.

One consequence of the equivalence principle is that the path of a light beam will be bent by gravity. Imagine a photon crossing the box as it accelerates through space. As the photon is crossing from one side to the other, the floor of the box is accelerating upwards, which means that the photon appears to fall downwards. Because a gravitational field is equivalent to acceleration, the same must also hold true in that case.

A second consequence is that time slows in a gravitational field. This effect, called gravitational time dilation, means that observers at different distances from a large object (which produces a gravitational field) will obtain different measurements for the time elapsed between two events. This is a direct consequence of the fact that an observer outside the box – that is, one outside the gravitational field – sees the photon follow a straight path, but the person in the box sees it follow a longer, curved path. Because the speed of light cannot change, the clock in the box has to run slower to allow both journeys to be made in the same time.

SPACETIME CURVES

In everyday terms we are used to dealing with the geometry of flat plane surfaces. One of the things we accept is that parallel lines will never meet. The surface of a

Einstein described gravity, not as a force, but as a consequence of the bending of spacetime by the matter embedded in it.

sphere is a flat surface, but it is also curved and the laws of geometry become a little different. There are no straight lines on a curved surface, but we can construct lines that are as straight as possible. Mathematicians call these 'geodesics'.

Einstein took this property of curvature and drew a comparison with the way gravity works. For a very small region of spacetime – for instance, a uniformly moving box floating freely in space – there is no gravity and the laws of the spacetime of special relativity hold sway. If we now add gravity to the situation – for instance, by placing a large planet in the path of the box – the box accelerates as it feels the force of gravity and its path begins to curve towards the planet.

Einstein looks at things in a different way. Rather than exerting a force, a mass causes a distortion of spacetime. Empty spacetime, the spacetime of special relativity, is flat. But matter causes spacetime to curve. Just as there are no straight lines on the surface of a sphere, there are no straight lines in curved spacetime. The planet, rather than exerting a force, has redefined the geometry of spacetime. The closest we can get to the straight line in curved spacetime, just as on a sphere, is a geodesic, a curve that is as straight as possible. The box heading towards the planet has not been deflected from its previous straight-line course, rather the distortion of spacetime has changed the form a straight line can take.

The earth makes a dent in spacetime, curving it around itself. The moon follows a straight path through the earth-curved spacetime, which to us appears to take it on a circular orbit (as the moon curves spacetime too, its path is really elliptical) around the earth. General relativity predicts that light rays will be bent by gravitational fields because light also follows geodesics through spacetime. This bending of light by gravity was one of the first confirmations that Einstein's theory was right.

This is the basis of Einstein's theory of general relativity. Newton's gravity is a force that acts on objects and influences their movement, but gravity in Einstein's universe is the result of curved spacetime. Objects still follow the straightest possible paths through spacetime, but because spacetime is now curved, they accelerate as if they were under the influence of a gravitational force.

In Einstein's universe, matter and spacetime interact in a complex and ever-changing dance. Matter distorts the geometry of spacetime and this distorted geometry dictates how matter moves through it. As the matter moves and the sources of gravity change positions, so the swirling curves of spacetime ebb and flow. As physicist John Archibald Wheeler succinctly summarized it: 'Spacetime tells matter how to move; matter tells spacetime how to curve.'

42

The end of spacetime

Black holes

In 1783, British clergyman John Michell posed an interesting question. Could there be an object so massive, with an escape velocity (the speed an object has to reach to escape the gravitational influence of another) so high, that light itself would not be able to escape? Michell called these 'dark stars' and speculated that, although invisible, they could be detected by their gravitational influence on other objects.

General relativity predicts the existence of regions where the density of matter is so intense that it warps and curves spacetime to such an extent that not even light has a spacetime path it can follow to escape it. In 1916, the year after Einstein published the theory of general relativity, German physicist Karl Schwarzschild used Einstein's equations to calculate what became known as the Schwarzschild radius. This described how far a given mass would have to be compressed in order to create a gravitational field powerful enough to trap light. The size of the Schwarzschild radius is proportional to the mass of the object.

Every galaxy may have a black hole at its centre.

THE EVENT HORIZON

The Schwarzschild radius defines the border that separates the black hole from the rest of the universe. This boundary, also known as the event horizon, is a one-way door – matter or energy can cross over it from the outside, but never comes out again. Because spacetime is so severely distorted, strange effects occur at the event horizon of a black hole. An observer watching someone fall towards the event horizon would see their clocks running slower and slower until, at the event horizon itself, time appeared to freeze. For the person falling, the opposite would be true – they would see time in the rest of the universe speed up and perhaps even witness its end before crossing the event horizon.

In 1928, the Indian astronomer Subrahmanyan Chandrasekhar calculated that when a star around 1.4 times more massive than the sun exhausted its thermonuclear fuel supply, the power of its gravitational force would be greater than its atomic particles could resist. The star would simply go on collapsing until it formed a singularity.

A singularity is a point where some otherwise measurable property becomes infinite. For example, the density of the material at the centre of a black hole is infinite, because the mass of the star has been compressed into zero volume under the pull of infinite gravity. At the centre of a black hole, spacetime has infinite curvature and space and time cease to exist in any meaningful sense. The laws of physics as we know them break down in the singularity, including relativity. Singularities were anathema to Einstein, who believed that there was no place for such infinities in a proper

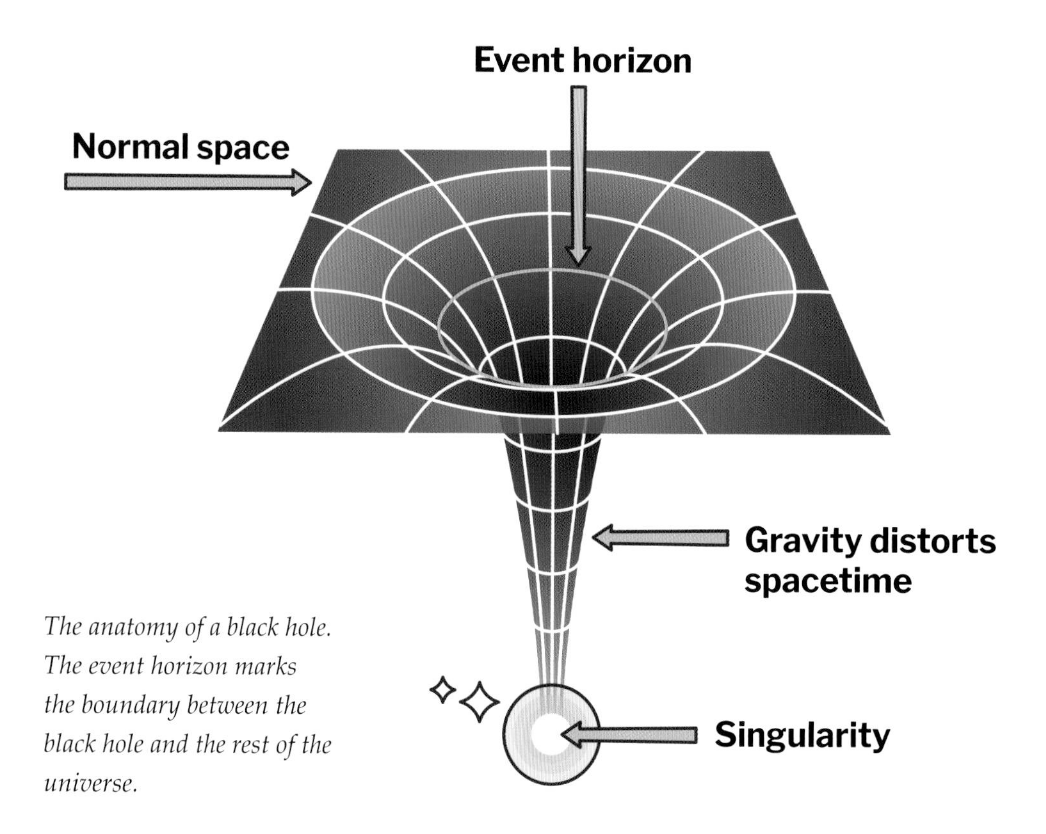

The anatomy of a black hole. The event horizon marks the boundary between the black hole and the rest of the universe.

mathematical description of the universe and that black holes would not form.

SEEING A BLACK HOLE

While there was no shortage of theories and speculations as to the nature of black holes, actually finding one proved more elusive. Astronomers need to look for indirect ways to deduce the presence of a black hole, for example by looking for unexpected movements in nearby stars. If a star that was part of a binary system collapses into a black hole it may start to pull gases in towards itself from the outer layers of its neighbouring star. This gas swirls in around the black hole, forming an accretion disc that reaches such high temperatures that it emits X-rays.

In 1971, astronomers suspected that an unusual source of x-rays in the Cygnus constellation, dubbed Cygnus X-1, resulted from a bright blue star in the region being ripped apart by a large, but invisible, object. Further research confirmed that Cygnus X-1 was indeed a black hole, 21 times more massive than the sun.

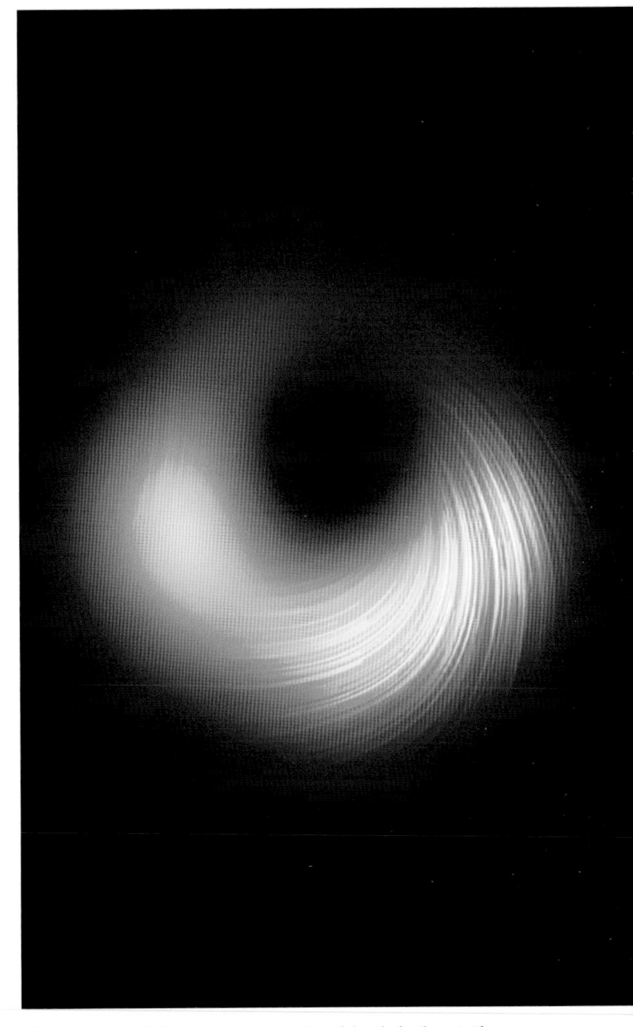

An image of the supermassive black hole at the heart of the M87 galaxy 55 million light years away. It is estimated to have a mass billions of times greater than that of the sun.

Radio galaxies shoot out beams of highly energetic particles on a vast scale. When these beams interact with clouds of intergalactic gas they cause them to emit radio waves which can be detected by telescopes on earth. There was only one possible power source: matter falling towards a compact mass and forming a high-energy accretion disc. Astronomers now believe that supermassive black holes with more than a million times the mass of the sun lie at the heart of most likely every galaxy, even our own Milky Way. The largest black hole currently known is the frankly incomprehensibly vast TON 618, located some 18.2 billion light years from earth, which has a mass equal to 66 billion times that of the sun.

43

A far away cataclysm

Detecting gravitational waves

Gravitational waves are like ripples in spacetime caused by particularly energetic disturbances. Einstein's equations showed that cataclysmic events, such as two black holes colliding or a massive supernova explosion, would be like large rocks being dropped into the pond of spacetime, sending out waves of distorted space across the universe at the speed of light.

As early as 1893, British physicist Oliver Heaviside speculated that gravity was transmitted as waves through a gravitational field, similar to the movement of electromagnetic waves. French physicist Henri Poincaré postulated in 1905 that gravity was transmitted at the speed of light in the form of waves, which he called *ondes gravifique*. Einstein wasn't so sure, at first suggesting that there were no gravitational waves, but later admitting that though they might exist there was no way they could ever be measured.

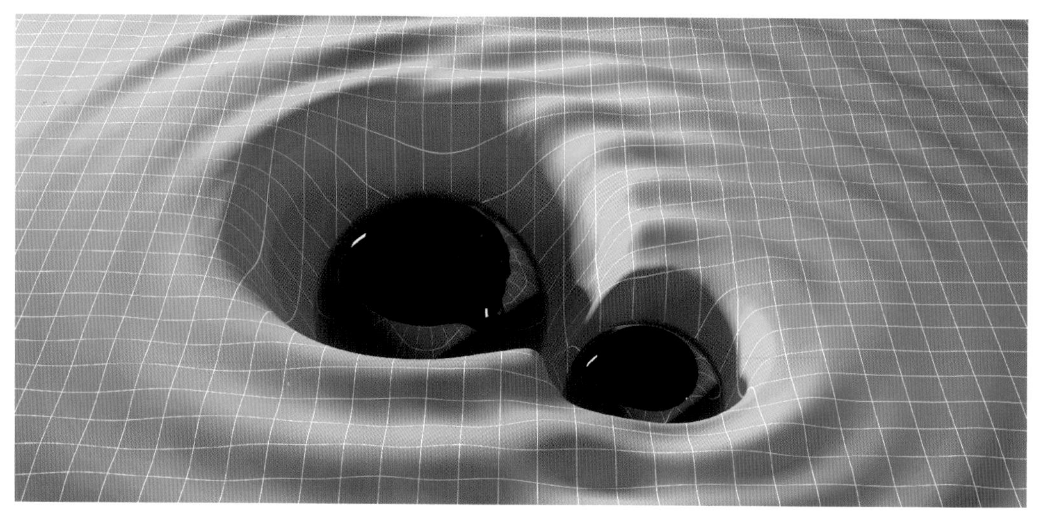

A cataclysmic event, such as the merging of two black holes, sends gravitational waves rippling across spacetime.

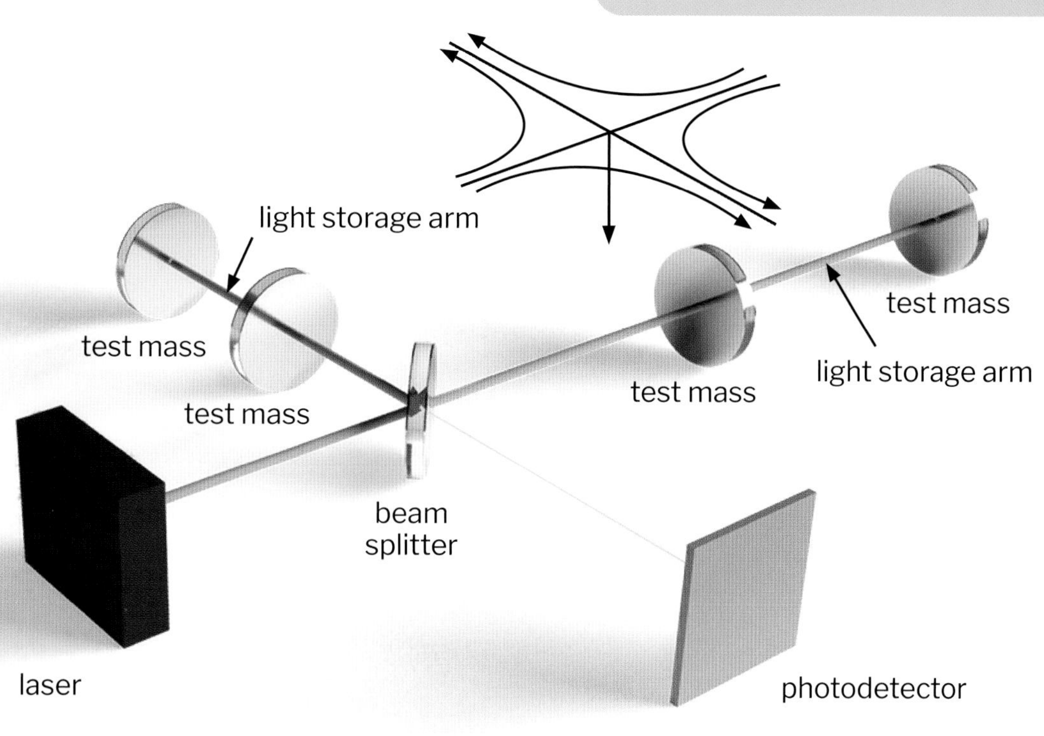

A schematic diagram of the LIGO interferometer depicting the gravitational wave arriving directly above the detector.

The equations of general relativity are extraordinarily difficult to solve, but by the late 1950s further work by a number of physicists had established that general relativity theory did indeed predict gravitational waves as a real phenomenon. There was still no means of detecting them, however.

In 1974, astronomers at the Arecibo Radio Observatory in Puerto Rico discovered a binary pulsar – two extremely dense and heavy stars in orbit around each other. Knowing that this system could be used to test Einstein's prediction, the astronomers began making careful observations of the system. Eight years of meticulous data-gathering revealed that the pulsars were getting closer to each other at exactly the rate general relativity predicted that they would. Another 40 years of close monitoring of the observed changes in the orbits of the pulsars was in such close agreement with general relativity that the researchers had no doubt that the binary system was emitting gravitational waves.

All of the confirmations of the existence of gravitational waves had been indirect or determined mathematically, and not through actual physical proof. This changed in 2015 when the Laser Interferometer Gravitational-Wave Observatory (LIGO) here on earth detected gravitational waves for the first time. The waves it detected were generated by two colliding black holes nearly 1.3 billion light years away. Although they were the result of such an extremely violent event, by the time the waves detected by LIGO reached the earth they were no bigger than the width of a proton.

LIGO is a triumph of engineering skill and ingenuity. It consists of two L-shaped detectors built 3,000 km (2,000 miles) apart, one in Washington State and the other in Louisiana in the United States. Other detectors in Italy, Germany and Japan help to further pinpoint the accuracy of the readings. The LIGO detectors have two right-angled arms each 4 km (2.5 miles) long, housed in vacuum chambers with a mirror at each end. A laser beam is split into two and sent down each arm. The beams bounce back from the mirrors and recombine at a light detector, creating an interference pattern. The interferometer is carefully set up so that, normally, the recombined beams cancel each other out. A gravitational wave stretches one arm and compresses the other, minutely changing the distance each laser beam travels and so altering the interference pattern when they recombine. Working in unison, the detectors can measure a motion 10,000 times smaller than an atomic nucleus – equivalent to measuring the distance to the nearest star to a precision smaller than the width of a human hair.

An aerial view of the LIGO interferometer site at Livingston, Louisiana, USA.

44

Star power

Nuclear fusion

Almost all of the energy available to us on earth can be traced back to the sun (the only exception being that produced by radioactive elements). In common with other stars, the sun produces energy through the process of nuclear fusion, forcing smaller atomic nuclei together to form larger ones.

Most stars produce energy by converting hydrogen nuclei (single protons) into helium nuclei (two protons and two neutrons). Getting two positively charged protons to fuse together is not an easy matter. It requires overcoming the powerful force of electrostatic repulsion, called the Coulomb force, pushing them apart.

The protons moving around in the core of the sun have tremendous energy; at temperatures in excess of 15 million °C (27 million °F) each proton travels at around 500 km/s (311 miles/sec). Collision between protons happens extremely frequently, in the order of billions of times each second. It would take only a tiny fraction of these collisions to result in the formation of deuterium, the first stage in the chain reaction leading to helium, to produce the colossal amount of energy pumped out by the sun. But none of them do. Even temperatures of 15 million °C aren't enough to overcome the Coulomb force. Nuclear fusion only takes place at all thanks to a combination of quantum mechanics and the weak nuclear force.

PARTICLES AS WAVES

Protons aren't simply particles. According to quantum mechanics they are waves as well. Each proton is a quantum, described by a wavefunction that describes its location as a probability. It is the overlapping wavefunctions of interacting particles that allows them to interact, even when the Coulomb force would otherwise force them apart. When the quantum wave hits the energy barrier of the Coulomb force there is a small, but non-zero probability that the proton will be located on the other side of the barrier. The proton has not surmounted the barrier, but rather gone through it, a phenomenon known as quantum tunnelling.

Quantum tunnelling

Imagine throwing a ball against a wall and seeing it disappear through to the other side instead of bouncing back. The phenomenon of quantum tunnelling allows electrons and other particles to do something very similar, passing through barriers that seem impassable. This oddity arises from considering electrons, for example, as stretched-out waves of probability rather than particles existing at a particular point. The Heisenberg uncertainty principle forbids us from knowing how much energy a particle has, or its exact location at any precise moment in time. There is a chance, albeit a very small chance, that the electron's probability wave will extend to the other side of the barrier. And sometimes it does. The effect is seen in transistors in which quantum tunnelling allows electrons to pass across a junction between semiconductors.

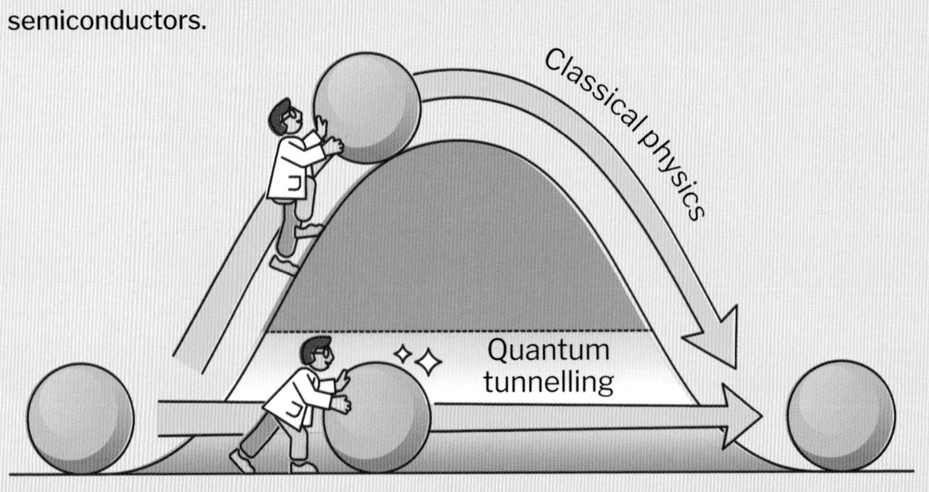

Quantum tunnelling allows a particle to cross a barrier it would not normally have sufficient energy to surmount.

Even then there are further difficulties to overcome. A deuterium nucleus isn't formed from two protons, but a proton and a neutron. This means that one of the protons has to be converted into a neutron by changing one of its up quarks into a down quark, which happens because of the action of the weak nuclear force. The process by which a proton changes into a neutron is called beta decay.

The odds of these events happening are ridiculously small, even longer than a jackpot win on the lottery, but it is just enough to ensure that fusion takes place on a scale sufficient to make a star shine.

WE HAVE IGNITION

Towards the end of 2022 researchers made a breakthrough in the quest to harness nuclear fusion as a clean source of energy. The National Ignition Facility (NIF) at Lawrence Livermore National Laboratory in California managed to release 2.5 megajoules of energy after using just 2.05 megajoules to initiate the reaction, a phenomenon known as ignition – creating a nuclear reaction that generates more energy than it consumes. A set of 192 lasers were fired at a pea-sized gold cylinder containing a frozen pellet of the hydrogen isotopes deuterium and tritium. The laser pulse raised the capsule temperature to stellar levels and the hydrogen isotopes fused into helium in a cascade of fusion reactions. Making fusion power a viable energy source is still some way off, but these are vital steps along the way.

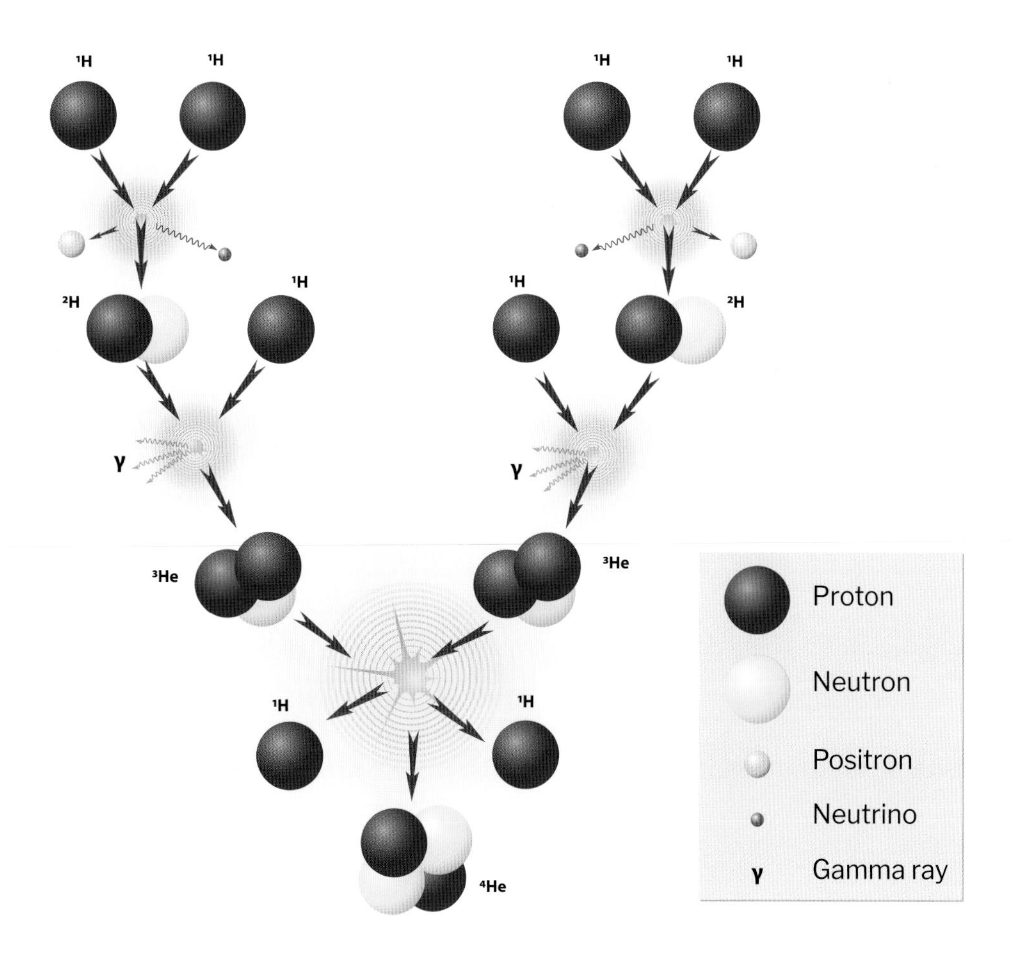

The fusion chain reaction that leads from hydrogen to helium.

45

The grand scale

The shape of the universe

Since the beginning of time people have looked up at the sky and wondered about what they saw there. We have come a long way from the ancient belief that the earth was at the centre of a universe bounded by a dome of stars. Today, we are forming ideas about the beginning and end of time and the vast expanse of space stretching almost beyond imagination.

An almost literally earth-moving revolution in human thinking took place in 1543 when Nicolaus Copernicus rejected the prevailing earth-centred view of the universe in favour of a heliocentric perspective in which the earth moved round the sun. Before Copernicus, astronomers had accounted for the observed motions of the stars and planets by imagining them being fixed to crystal spheres centred on the earth. Copernicus still believed in these nested perfect spheres; he just moved them to a new location to make the calculations easier.

PROBLEMS OF PERSPECTIVE

As astronomers got to grips with the new perspective other problems presented themselves. In 1692, Isaac Newton received a letter from the Reverend Richard Bentley. If the universe were infinite, as many supposed it to be, then surely, suggested the Reverend Bentley, every part of the universe should feel the pull of gravity, and therefore should it not collapse in on itself?

Newton tried to solve this conundrum by arguing that if the stars were evenly distributed in space, then the force of gravity would act equally in all directions and a balance would be maintained. He quickly realized, however, that the slightest movement of any star would upset the balance and his cosmic house of cards would fall.

Newton and Bentley's error was in believing the stars to be stationary. It was Edmund Halley, of comet fame, who first observed that a few stars had shifted from the positions recorded for them on Greek star maps. Halley posed another problem. If the universe were infinite, then wherever you looked there ought to be a star – the entire sky should be shining as brightly as the sun. Obviously,

If the universe is infinite in size, why isn't the night sky full of stars?

it wasn't – an observation that had led Johannes Kepler to conclude in 1610 that the universe couldn't be infinitely large. The problem became known as Olbers' Paradox, after the German astronomer Heinrich Olbers. He suggested in 1823 that there must be clouds of dust between the stars, hiding some of them from our view. But this solution was flawed also. Given long enough, the energy from the distant stars would heat up the gas clouds until they glowed, and the sky would be filled with light. The answer to the problem came a hundred years later.

At the beginning of the 20th century, evidence was beginning to accumulate that the universe was much, much bigger than anyone had previously imagined. Most people believed that the Milky Way galaxy was the whole universe but there was speculation about whether or not some objects in space might lie beyond it.

VAST – AND GETTING BIGGER

Edwin Hubble settled the argument in 1923 when he used the most powerful telescope in existence at the time, the Hooker Telescope at the Mount Wilson Observatory in California, to examine stars in the Andromeda nebula. He estimated their distance as 800,000 light years. Although this turned out to be an underestimate, it confirmed that Andromeda was a galaxy in its own right. Hubble went on to find other galaxies that were even more distant. A picture began to emerge of a universe vast beyond imagining, stretching over billions of light years, with a hundred billion galaxies each containing around a hundred billion stars. It turned out that we were a very long way indeed from being the centre of creation.

However big it turned out to be, one belief about the universe still held – that it was static. No one thought that it might expand or contract. In 1929, Hubble became famous overnight when he made another game-changing discovery. The light coming to us from the distant galaxies is shifted towards the red end of the electromagnetic spectrum, indicating that these galaxies are moving away from our solar system. The further they are away from us, the faster they are retreating. Those that are twice as far move away roughly twice as fast. The best explanation for this was that the universe is expanding.

Here was the answer to Olbers' Paradox. The light from the furthest reaches of the universe has not had time to reach us – and perhaps never will. The universe is dark because it started with a Big Bang.

What are the consequences of an expanding universe? The ultimate fate of the universe depends on the balance between the rate of expansion, expressed by a factor called the Hubble Constant, and the curving of spacetime by gravity, which is determined by the amount of matter in the universe. There are three possibilities. First, the amount of matter

The universe is believed to be expanding like a balloon.

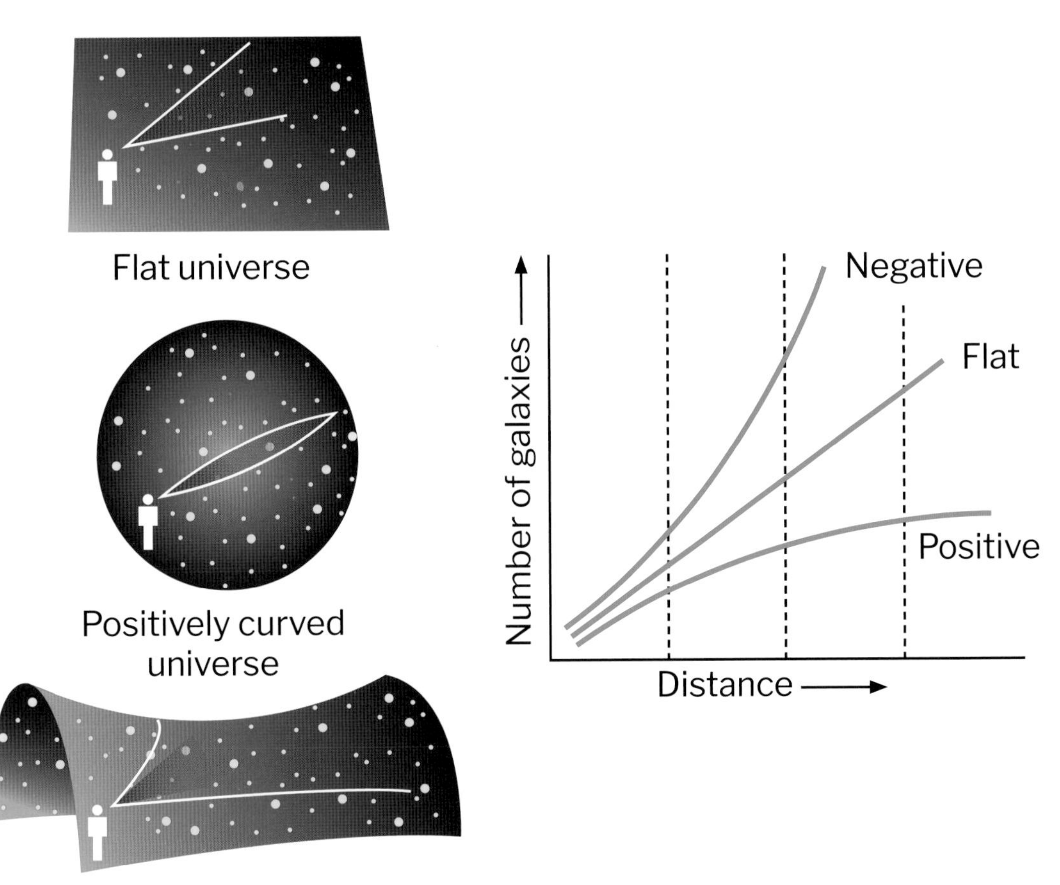

The curvature of the universe, and perhaps its ultimate fate, is determined by the amount of matter in it.

in the universe exceeds what is known as the 'critical density'. This isn't a great deal – roughly six hydrogen atoms per cubic metre. In this case, the expansion is slowed down, stopped and reversed by gravity and the universe eventually collapses in a Big Crunch. Second, the density of the universe is a little less than the critical density. The expansion continues but at a slower and slower rate. Third, and mysteriously this appears to be what actually is happening, the rate of expansion accelerates.

The amount of matter in the universe also determines its geometry. If the density of the universe is greater than the critical density, the geometry of space is closed and curved like the surface of a sphere. If less than the critical density, the geometry is open (infinite), and curved like the surface of a saddle. If the density exactly equals the critical density, then the geometry of the universe is flat like a sheet of paper, and infinite in extent.

46

Where it begins

Big Bang theory

Where did the universe come from? If we take the expanding universe and rewind the clock, throwing it into reverse, all matter, all energy, all of space and all of time contract down into a singularity of infinite density and gravity and zero size. As best we know, everything that makes up the universe today expanded out from this zero point in an event that came to be known as the Big Bang.

Towards the end of the 1920s, Belgian priest and astronomer Georges Henri Lemaître put forward his idea of a 'primordial atom'. Lemaître deduced the fact that the speed with which the most distant galaxies appear to be receding from us is proportional to their distance. He spoke of a 'day without a yesterday' and suggested that in the distant past all the mass in the universe had been concentrated into a single super atom. According to Lemaître, this primordial atom divided again and again, eventually giving rise to all the matter we see today.

Belgian astronomer Georges Henri Lemaître proposed that all the mass in the universe had once been condensed into a single 'super atom'.

After the Second World War, Ralph Alpher and George Gamow suggested that, in the beginning, the universe was formed from a hot soup of atomic particles at a temperature of trillions of degrees. This cooled as it expanded and its energy was spread over greater and greater volumes of space. Their calculations suggested that hydrogen and helium should be by far the most common elements and that there should be ten atoms of hydrogen for every one atom of helium. This was exactly the ratio that had been determined by astronomers.

In 1948, Alpher predicted that radiation from the early beginnings of the universe should still be detectable. This 'cosmic background radiation', as it was called, the last faint glow of the birth of the universe, would have a temperature of about -268°C (-450°F). Unfortunately, the equipment didn't exist at the time to refute or verify the theory.

A great many astronomers, among them the eminent British astronomer Fred Hoyle, refused to accept that the universe actually had a beginning. In 1950, Hoyle gave a radio talk on the subject. Hoyle was a fierce opponent of the expanding universe theory, preferring his own theory, which he called 'Steady State', in which matter is continually created and the universe remains much as it always has been. He referred to the rival theory as a 'big bang'. The term stuck, and from then on the idea that the universe started from an initial point became the 'Big Bang' theory.

STOP THAT PIGEON

In 1964, US radio astronomers Arno Penzias and Robert Wilson made a discovery that finally tipped the scales in favour of the Big Bang. While testing an astronomical microwave detector,

Arno Penzias and Robert Wilson were the first to detect the cosmic background radiation, the faint energy echoes of the Big Bang.

they were concerned to discover that the device seemed to be picking up noise from all over the sky. At first, they thought that pigeon droppings might be causing it to malfunction, but after cleaning out the detector – and shooting the pigeons – they discovered that the noise was coming from outside the atmosphere, and from every direction. Whatever time of day they tried, it never varied. It was confirmed that the mysterious signals were the cosmic background radiation Alpher had predicted and therefore proof of the Big Bang.

According to general relativity, the beginning of the universe could have been in a Big Bang, but should it? That's the question physicist Roger Penrose posed in 1965, when he demonstrated mathematically that a star collapsing under its own gravity would eventually form a black hole singularity, becoming trapped in a region of space that shrinks to zero in which the curvature of spacetime is infinite. Stephen Hawking read Penrose's work and realized that

An artist's impression of the Big Bang.

Not a bang at all

The Big Bang was not a sudden explosion of all the matter in the universe out into space. Space, time and everything else came into existence with the Big Bang. Before the Big Bang there was no space for anything to explode into. There is no 'centre' of the universe from which everything expanded and no cosmic vantage point from which it would be possible to watch the Big Bang unfold. The Big Bang was an eruption of space and time that carried all the mass and energy of the universe along with it. It is meaningless and ultimately unanswerable to speculate about what the universe is expanding into.

by reversing the direction of time, so instead of collapsing to zero, there is an expansion out from zero, the theorem still held. In 1970, Penrose and Hawking produced a joint paper which offered mathematical proof that if the description of the universe given by Einstein's theory of general relativity was correct, and the universe contains as much material as we observe that it does, it must have begun with a singularity.

Current calculations date the Big Bang to approximately 13.8 billion years ago. The furthest reaches of theory takes us back to the Planck Epoch, a period so close to the beginning of the universe that our ideas of physical laws break down. The Planck Epoch took place just one 10-millionth of a trillionth of a trillionth of a trillionth of a second (10^{-43} seconds) after the Big Bang. The entire universe was just 10^{-35} metres across with a temperature of 10^{32}°C. The four fundamental forces – gravity, the electromagnetic force and the strong and weak nuclear forces – are believed to have been a single unified force at this time. Figuring out what was happening during the Planck Epoch requires a theory of quantum gravity, uniting general relativity with quantum mechanics. Somehow, we would need to describe the quantum state of the entire universe. This is an endeavour that has so far defeated the best efforts of physicists.

Following the Planck Epoch comes the Grand Unification Epoch during which gravity separated out from the other forces. This lasted from 10^{-43} seconds to 10^{-36} seconds.

Between 10^{-36} seconds to 10^{-32} seconds after the Big Bang the universe enters the Inflationary Epoch. Within this tiny fraction of a second the universe grew 10^{26} times bigger, swelling from about the width of a molecule to several light years across. Although objects within spacetime are limited by the speed of light it was the structure of spacetime itself that expanded at this extraordinary rate. What powered this event is a matter

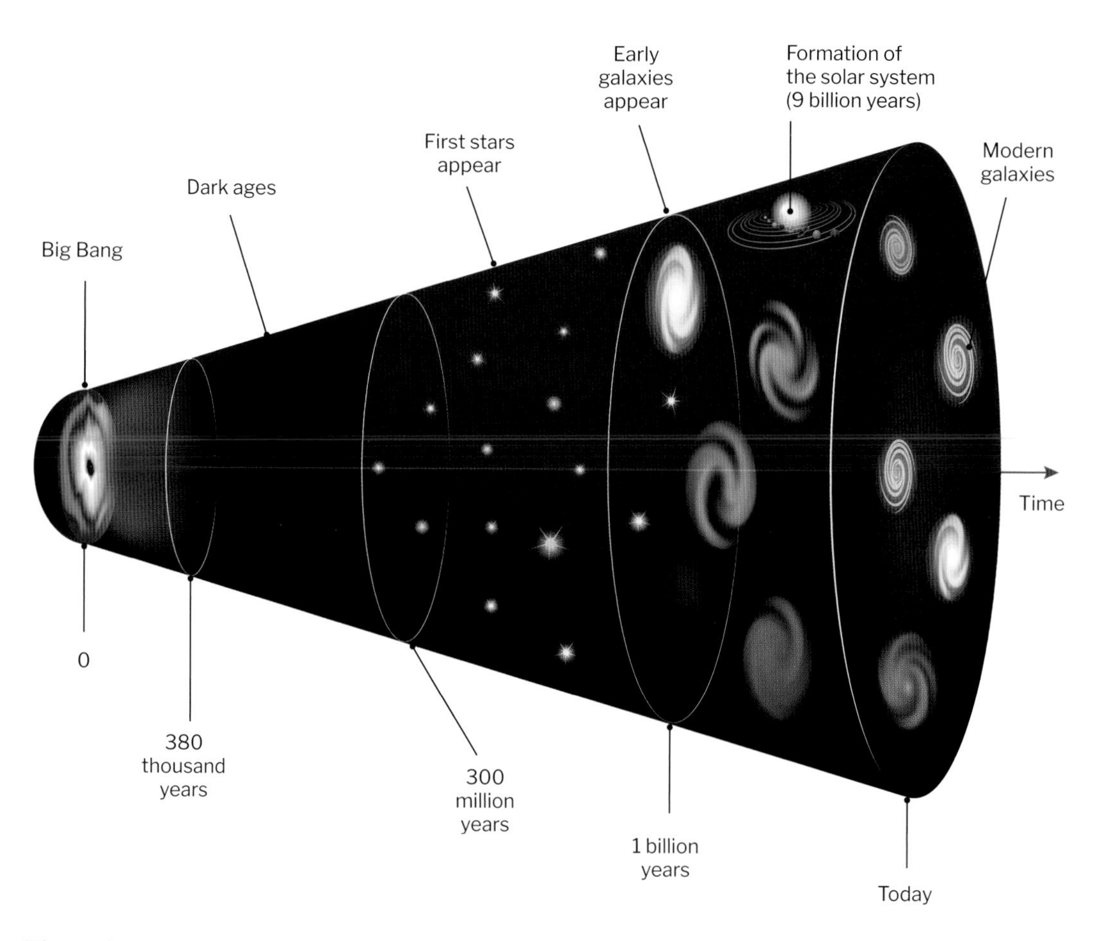

The evolution of the universe, from the Big Bang to the present.

of speculation, but one possibility is that it was triggered by the separation of the strong nuclear force with the release of a huge amount of energy.

Following the Inflationary Epoch the universe entered the Electroweak Epoch during which the electromagnetic and weak nuclear forces are still combined. A few millionths of a second after the Big Bang, the four fundamental forces have become distinct from each other, and quarks joined together to produce protons and neutrons. Within minutes, these protons and neutrons had combined into nuclei. It took another 380,000 years or thereabouts for electrons to be trapped in orbits around nuclei, forming the first atoms – mainly hydrogen and helium. The first stars may have formed from clouds of gas around 150–200 million years after the Big Bang.

47

Sight unseen

Dark matter

Astronomers were perplexed to discover that galaxies were doing something that seemed impossible. They are spinning so fast that the matter in them couldn't possibly be generating enough gravitational force to hold them together. They should be flying apart. Why weren't they?

In 1933, Swiss astronomer Fritz Zwicky discovered that the actual mass of the galaxies in the Coma Cluster must be far greater than the observable mass if the cluster wasn't to fly apart. There had to be some other factor at work that wasn't showing up in any observation. American astronomer Vera Rubin reinforced the evidence for this 'invisible' matter when she showed that spiral galaxies were rotating too quickly to be held together by visible matter. Astronomers and physicists concluded that something undetectable was providing galaxies with the extra mass needed to generate the gravity to keep then intact. Zwicky called this strange and mysterious matter d*unkle materie* – 'dark matter'.

Swiss astronomer Fritz Zwicky, who discovered the first indications that there might be more to the universe than we could see.

Dark matter appears to spread across the cosmos in a net-like pattern, with galaxy clusters forming at the nodes where strands of dark matter intersect. It seems to outweigh visible matter by a ratio of roughly six to one. As far as we can tell, dark matter does not interact with the electromagnetic force. Because it does not appear to absorb, reflect or emit electromagnetic radiation, it seems impossible to detect with any of the observational tools we normally use, whether radio, infrared, optical or x-ray telescopes. Dark matter's evident gravitational effect is the only proof we have for its existence. But what is dark matter?

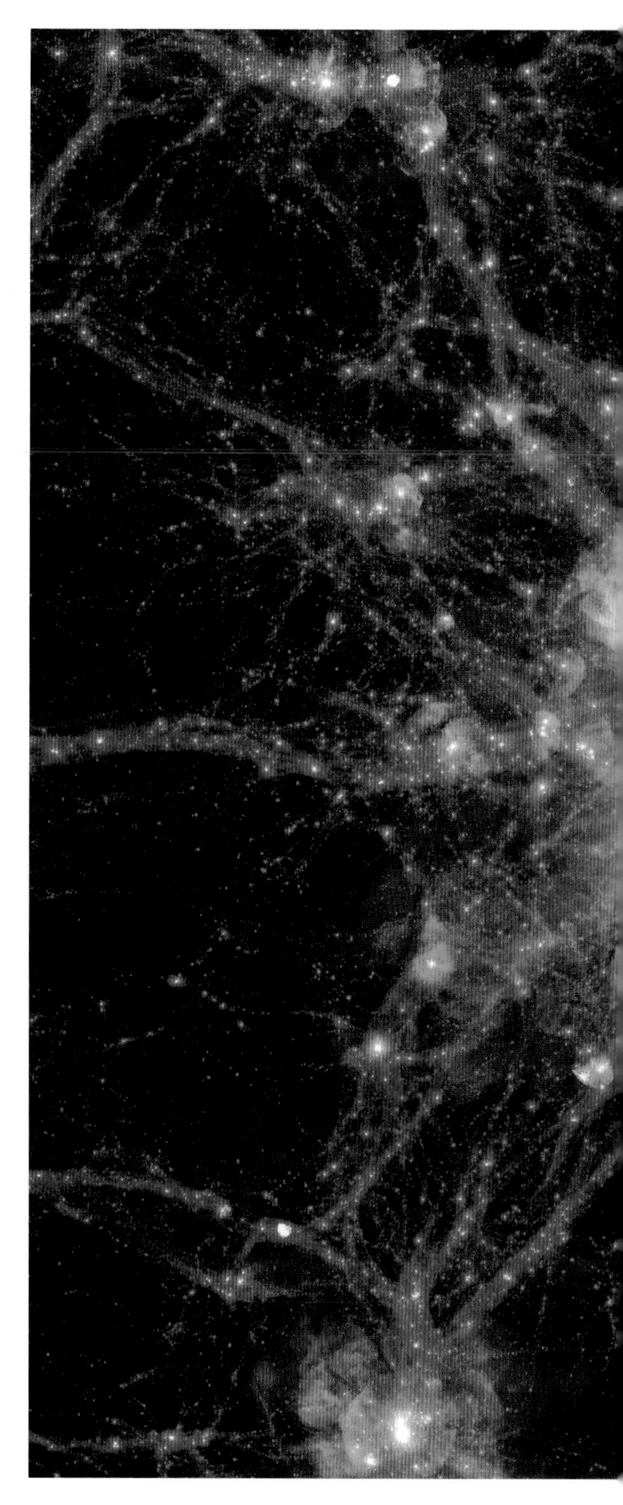

DARK MATTER CANDIDATES

It is perhaps easier to say what it isn't rather than what it is. It doesn't take the form of visible matter – it isn't stars and planets we haven't yet discovered. It isn't made up of dark clouds of normal matter. We know this because we would be able to detect these clouds as they absorbed radiation passing through them. It is not antimatter, because we see no evidence of the gamma rays that are produced when antimatter annihilates with matter. We can also rule out giant undiscovered black holes. Light is bent by the intense gravity of a black hole but again there is no evidence of this happening on a large enough scale to account for the missing matter in the universe.

Most scientists think that dark matter must be composed of a different kind of matter to the protons, neutrons and electrons we are used to. One possible candidate is the so-called weakly interacting massive particles, or WIMPS. Believed to be ten to a hundred times more massive than a proton, they are difficult to detect because they do not interact with 'normal' matter particles. Another idea is that dark matter could contain supersymmetric particles, theoretical particles that are partners to the familiar particles that are included in the Standard Model of particle physics. One theory suggests the existence of

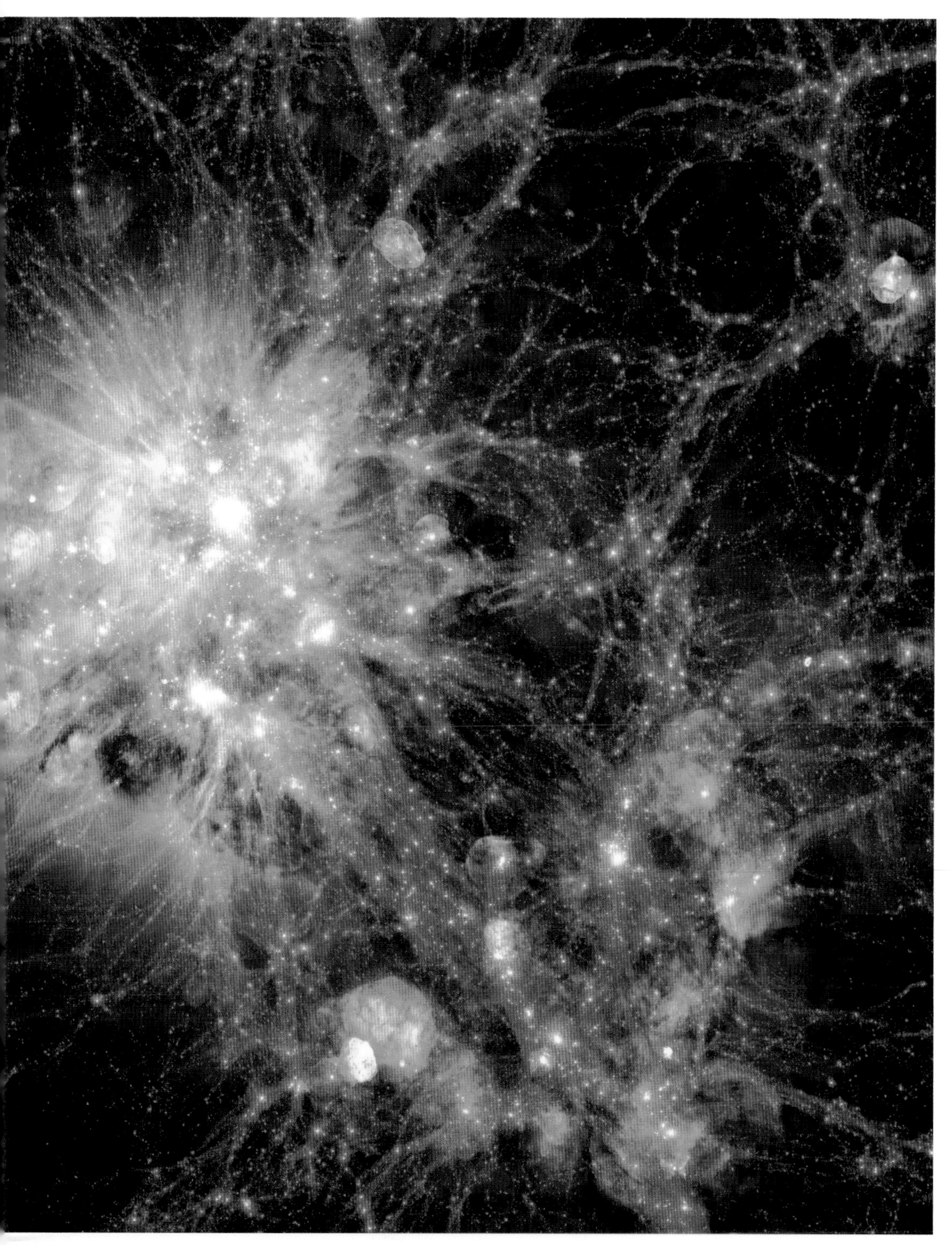

Invisible filaments of dark matter are thought to stretch across space.

Images from the Hubble Space Telescope have been used to study the distribution of dark matter in distant galaxy clusters.

a parallel universe of dark matter that has little in common with our 'normal matter' universe.

Many theorists believe that it should be possible to produce dark matter particles at CERN's Large Hadron Collider. The problem is, of course, if they were created at the LHC, they would almost certainly escape capture by the LHC's detectors, but their existence could be inferred from the amount of energy and momentum 'missing' after a collision.

48

Dancing in the dark

Dark energy

As if it wasn't perplexing enough to discover that the universe was filled with some unseen, unknown form of dark matter that outnumbered ordinary matter by a ratio of five to one, scientists have also found that the fate of the universe may be determined by the existence of a mysterious dark energy.

Astronomers in the early 1990s were fairly certain that the expansion of the universe following the Big Bang would either effectively go on for ever at an increasingly slower rate as gravity slowed the rate of expansion down, or the gravitational attraction would be sufficient to reverse the expansion and the universe would collapse back in on itself. There was a surprise in store at the end of the decade when observations of distant supernovae by the Hubble Space Telescope showed that, a long time ago, the rate at which the universe is expanding was actually slower than it is today. The expansion of the universe has not been slowing at all, as everyone believed it to be – it has been accelerating. No one knew how to explain this.

Albert Einstein's general theory allowed for the notion that the universe could be either expanding or contracting. In fact, the equations of general relativity don't allow for a static universe – if the universe was neither static nor collapsing, then it had to be expanding. In 1917 he introduced a term called the cosmological constant into his equations. This was a repulsive force that counterbalanced the attraction of gravity, keeping the universe static. Edwin Hubble's discovery of the redshifting of distant galaxies demonstrated that the universe really was expanding. Einstein removed the cosmological constant from his equations, glad that it wasn't necessary after all.

Many of today's theorists support the idea that the mysterious dark energy pushing the universe apart behaves very much like Einstein's discarded cosmological constant. One explanation for dark energy is that it is a property of space itself. Einstein discovered that it is possible for more space to come into existence – the expanding universe creates the space it expands into. General relativity theory predicts that 'empty

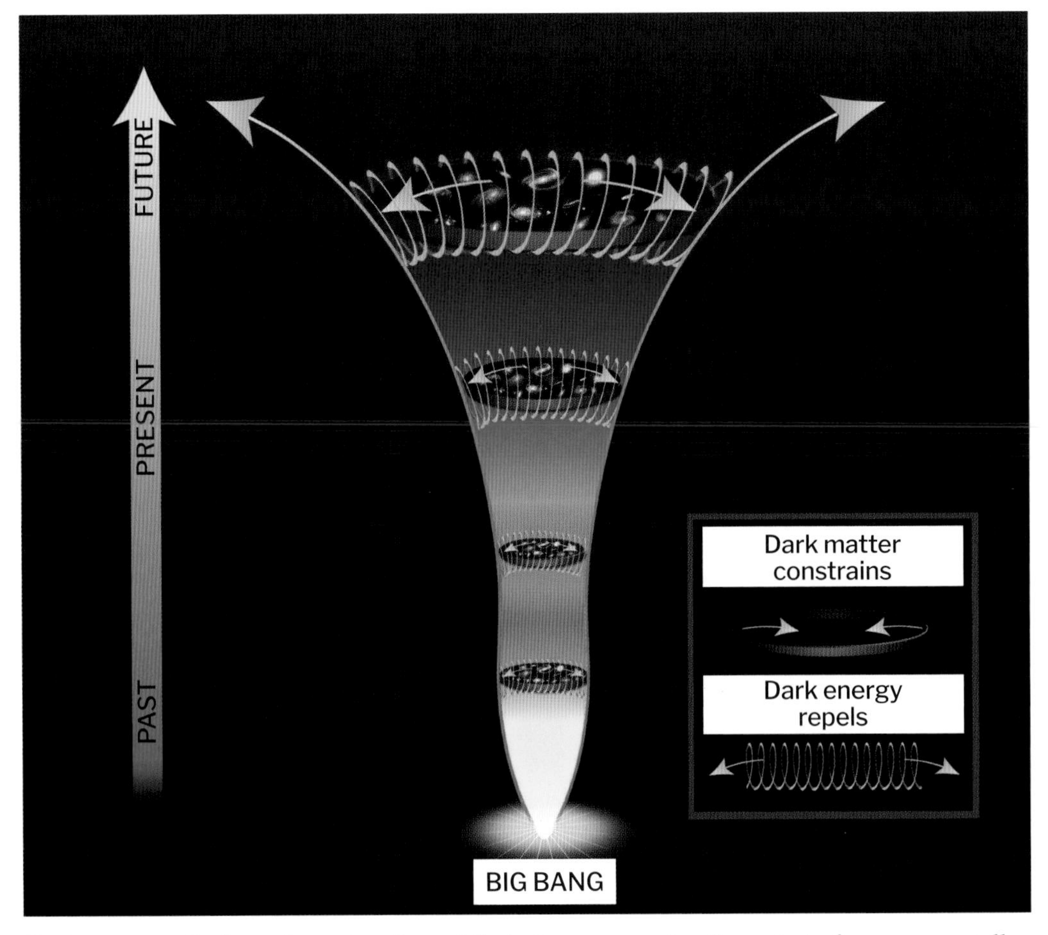

Dark matter pulls the universe together while dark energy pushes it apart. As far as we can tell, dark energy is winning.

space' can possess its own energy. As space comes into existence, more of this energy appears, causing the universe to continue expanding faster and faster. At the moment, no one understands why the cosmological constant has exactly the right value to cause the observed acceleration of the universe.

Quantum mechanics provides another possible explanation for dark energy. According to quantum theory 'empty' space is not empty at all. It is a seething mass of probabilities, full of temporary, or virtual, particles that continually blink in and out of existence. Unfortunately, when physicists tried to calculate how much energy this would impart to empty space, the answer came out very badly wrong. In fact, it came out 10^{120} times too big. That's a 1 with 120 zeros after it. This is not a trivial error.

Another possibility, which physicists find hard to accept, is that Einstein got it wrong and his theory of gravity is

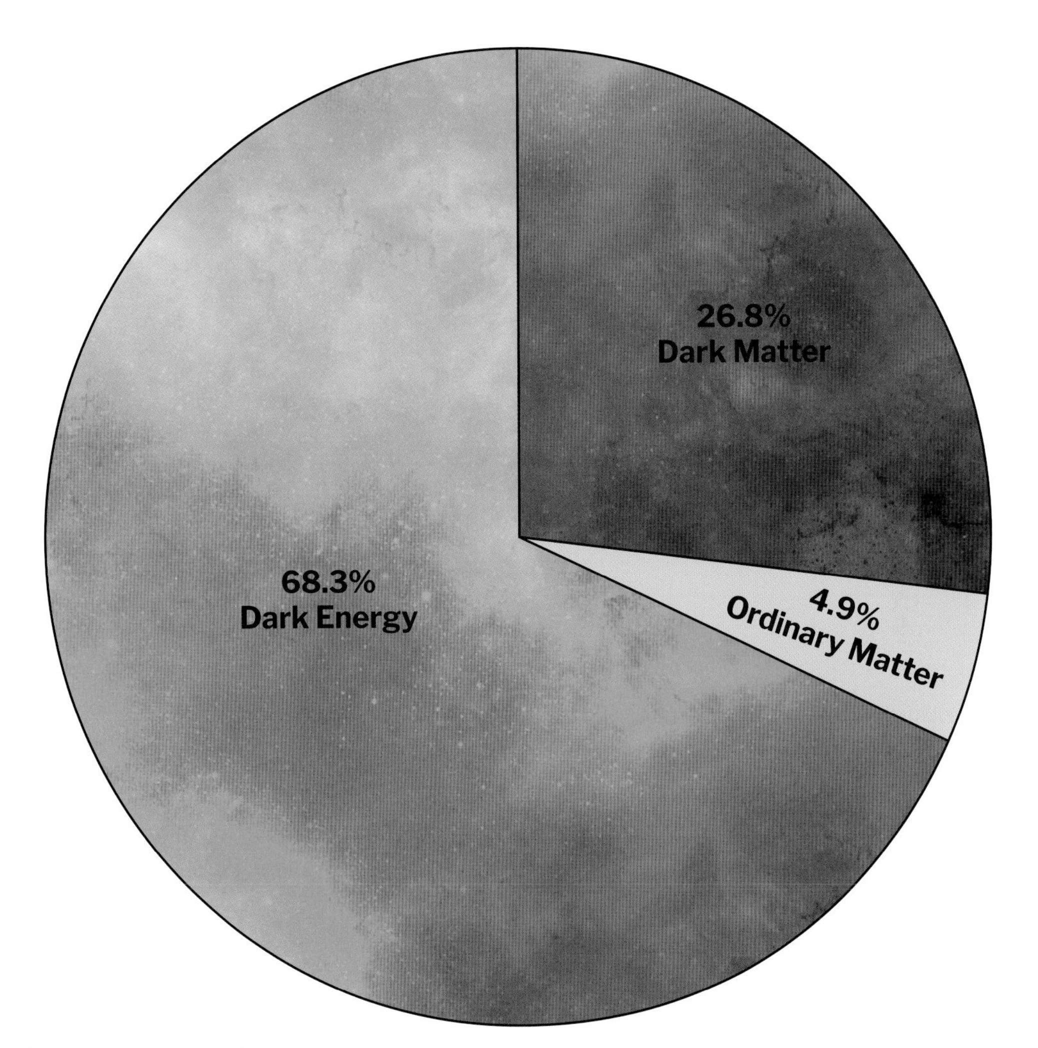

Dark energy and dark matter are by far the dominant factors in the make-up of the universe.

somehow flawed. If it turns out that a new theory of gravity is needed it will need someone who can be for Einstein what Einstein was for Newton. At the moment, there are no obvious candidates.

We know how much dark energy there is in the universe because we can measure how it affects the universe's expansion. It turns out that approximately 68 per cent of the universe is dark energy. Dark matter makes up about 27 per cent. The rest – everything we have ever seen or examined and formulated our theories of physics around, everything we have been pleased to call 'normal' matter – makes up less than 5 per cent of the universe. It really begs the question of what is 'normal'?

49

Across the great divide

Quantum gravity

Einstein's relativity theories provide a framework for understanding the large-scale universe of stars and galaxies; quantum mechanics describes how it works on the subatomic scale. These two theories are the twin pillars of modern physics. However, the two cannot be reconciled and there seems to be no easy way to unite the two.

The Standard Model of particle physics provides a description of the fundamental forces and particles that together describe the make-up of the universe and how it works. However, the Standard Model doesn't take into account Einstein's idea of spacetime and quantum mechanics cannot account for gravity. Of all the fundamental forces, gravity is the one that stands apart. Finding a way to fit gravity into quantum mechanics is a vital step in the search for a 'theory of everything' that would explain how the universe works on every level. Crucial to this is knowing whether or not gravity is a quantum phenomenon and that means establishing whether or not the theoretical force carrier of gravity, the graviton, exists or not.

According to quantum mechanics everything is made of quanta, or packets of energy, that can behave like both a particle and a wave. Detecting gravitons would prove gravity is quantum. The problem is that gravity, in comparison to the other fundamental forces, is extraordinarily weak. Physicist Freeman Dyson has estimated that if the LIGO detector that established the existence of gravitational waves were scaled up to detect gravitons it would become so massive that it would collapse in on itself to form a black hole.

THE CASIMIR EFFECT

One possibility is that the quantum nature of gravitons, should they exist, might offer a way to detect them. According to Heisenberg's uncertainty principle we can never accurately know a particle's position and momentum at the same time, only the probability of it being so. A consequence of this is that so-called empty space is never really empty. Virtual particles pop in and out of existence all the time. These ghost particles can take any form; they could even be gravitons.

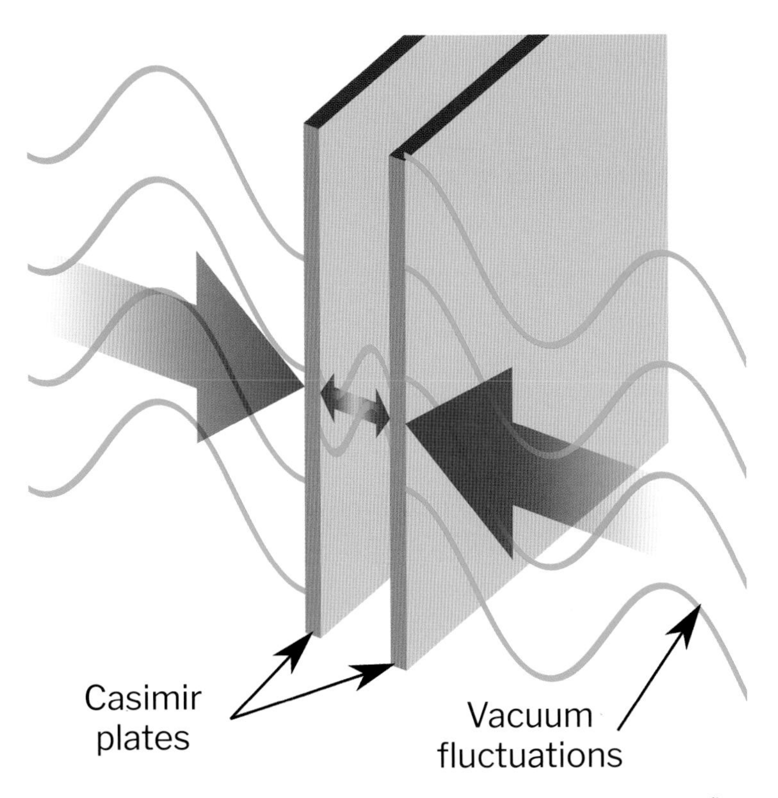

The Casimir effect is a small attractive force resulting from quantum vacuum fluctuations in the electromagnetic field.

Although these virtual particles are here and gone in an instant it has been established by experiment that they can generate detectable forces. One such is the Casimir effect, named after Dutch physicist Hendrik Casimir, who first predicted it in 1948. Virtual photons appearing and disappearing in a vacuum mean that it is full of fluctuating electromagnetic waves in all possible wavelengths. If mirrors are placed facing each other in a vacuum, some of the waves will fit between them but others will not. If fewer waves fit than those that don't the result is that the total amount of energy in the vacuum, between the plates is a bit less than the amount elsewhere in the vacuum resulting in the mirrors being pushed closer together. Research indicates that superconducting materials may reflect gravitons more strongly than ordinary matter so using two superconducting sheets in place of mirrors could possibly detect a gravitational Casimir effect. Attempts to do so have so far been unsuccessful, however.

Another possibility lies in studying the cosmic microwave background radiation, the faint afterglow of the Big Bang. When the universe expanded rapidly in the fraction of a second after the Big Bang, short wavelength quanta, including gravitons, would have become stretched out. This could theoretically leave detectable patterns of swirls in the background radiation, but these may be so weak they will never be detected.

LOOP QUANTUM GRAVITY

One approach to making gravity fit in to quantum theory is known as loop quantum gravity. This involves considering spacetime as being made up of tiny, discrete one-dimensional building blocks that lock together to make up the four-dimensional fabric of the universe rather than being a smooth continuum. The size of these spacetime fragments is of the order of the Planck length, approximately 10^{-35} metres, or about 23 orders of magnitude smaller than an atom.

So far none of the available theories provide a comprehensive description of quantum gravity and experimental evidence still seems out of reach.

Artist's impression of quantum space in loop quantum gravity.

50

World on a string

String theory and supersymmetry

One of the most hopeful candidates for a theory of everything is string theory. It not only provides a consistent description of the universe that unites all four fundamental forces and the fundamental particles of the Standard Model, but also offers a way to tie gravity into the overall structure as well.

The idea of string theory was first developed in 1969 by American physicist Leonard Susskind, who was attempting to explain how the strong nuclear force works to bind the atomic nucleus together. Since the early 20th century, nature's fundamental particles, such as electrons, quarks and neutrinos, have been portrayed as being objects with no internal structure. String theory challenges this. At the heart of string theory is the idea that the fundamental particles are really just different manifestations of one basic object: a string. Rather than being point-like, a particle is more akin to a one-dimensional vibrating filament of energy. The differences between one particle and another – their mass, charge and other properties – all depend on the vibrations of their internal strings. Like a skilled violinist conjuring a melody, nature manifests all the particles of the atomic realm through changes in the frequency of the subatomic string.

Of great interest is the fact that one of the 'notes' on the string corresponds to the graviton, the hypothetical particle that carries the force of gravity from one location to another, just as the photon does for the electromagnetic force.

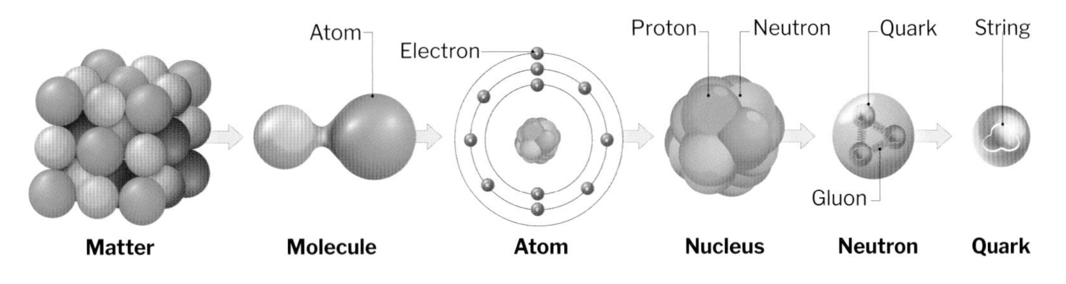

Is matter at the deepest level composed of vibrating strings?

This seemed to promise a solution to the problem of reconciling gravity and quantum mechanics.

An important feature of string theory is the idea of supersymmetry, championed by physicists such as Japanese-American Bunji Sakita and Italian Bruno Zumino, which postulates that every member of the Standard Model's particle zoo has a heavier twin. Quarks, for example, are partnered by 'squarks' (supersymmetric quarks). If supersymmetry theory is correct, then these super-twins may be the source of dark matter.

INTO THE MULTIVERSE

An additional complication of string theory is that its equations require that the universe has extra spatial dimensions beyond the familiar three to make them work. String theorists took up the idea first developed by Theodor Kaluza and Oskar Klein in the early years of the 20th century when they were trying to link Einstein's gravity with electromagnetism. Kaluza showed that if spacetime were extended to five dimensions, then four of those dimensions would encompass Einstein's general relativity equations, while the fifth dimension would be the equivalent for Maxwell's equations for electromagnetism. Klein later determined that the fifth dimension would be curled up so small that we would be unable to detect it. Perhaps there might also be additional tiny, compact dimensions wound up inside the 'normal' ones.

It has been suggested that because strings are so small they will vibrate not just in the 'big' dimensions that we move around in, but in the tiny unseen ones as well. Audaciously, some string theorists have predicted that since the vibrations of the strings determine the properties of the fundamental particles, properties we can detect experimentally, and the vibrations are determined by the shape of the extra dimensions, there might be a way to work back from our knowledge of the fundamental particles to a map of the esoteric dimensions.

Unfortunately, the number of mathematically allowable shapes for the extra dimensions runs into the billions. Susskind suggested that, if there was no one shape that was right, then perhaps they all were, each one the right shape within their own unique universe. Our familiar universe might be just one of a vast, perhaps infinite assembly of universes, each with features determined by the shape of their extra dimensions. 'Our' universe's hidden dimensions make possible the laws of physics that led to the existence of galaxies, stars, planets, the chemical elements, and life itself. In some other dimensional configuration different laws would apply and the universe would be a very different, probably lifeless place.

So, are strings 'real'? Proving their existence is problematic to say the least. The mathematics of string theory require them to be about a million billion times smaller than anything the world's most powerful particle accelerators have uncovered. It is possible that detectors such as the Super-Kamiokande in Japan may uncover evidence of proton decay, a currently unobserved phenomenon but one which is predicted by supersymmetry theory, but physicist Brian Greene believes we would need to build 'a collider the size of the galaxy' to have any hope of directly detecting strings. Some scientists believe the pursuit of strings is a futile endeavour because we will never be able to prove their existence. String theory has the potential to explain much of the workings of the universe and may open the way to an understanding of dark matter and dark energy but as yet there is no hard experimental evidence to back it up.

Are there extra dimensions beyond the ones we know?

Index

Picture credits

Alamy: 69, 78, 82, 90, 103, 110, 123, 139, 159, 186, 187

CalTech: 178 (Virgo Collaboration/Ligo)

CERN: 133

ESO: 193

Library of Congress: 98

NASA: 22, 170, 188, 194, 196

Science Photo Library: 44, 62, 118, 130, 137, 148, 176, 200

Shutterstock: 6, 8, 10, 13, 18, 20, 21, 28, 31, 32, 34, 36, 37, 38, 40, 42, 45, 46, 51, 60, 63, 67, 70, 71 (x2), 72, 73, 74, 76, 77, 79, 80, 81, 84, 85, 86, 89, 91, 92, 93, 94, 95, 97, 100, 114, 115, 120, 125, 126, 127, 128, 129, 143, 149, 151, 156, 157, 168, 171, 173, 174, 180, 181, 183, 184, 190, 201, 203

Wellcome Collection: 9, 33, 59, 105, 122

Wikimedia Commons: 15, 25, 26, 29, 35, 55, 56, 57, 66, 87, 107, 109, 111, 132, 135, 140, 141, 144, 146, 153, 166, 175, 177, 191, 199